This book is dedicated to my two family editors; my sister Linda and my very best friend Jean. Also, I would like to thank Darrell Urbien, the 3D expert at Orange Coast College, for refreshing my 3D knowledge.

Other workbooks written by
Cheryl R. Shrock:

Exercise Workbook for Beginning AutoCAD 2004, ISBN 0-8311-3198-5

Exercise Workbook for Advanced AutoCAD, ISBN 0-8311-3199-3

Exercise Workbook for Beginning AutoCAD 2002, ISBN 0-8311-3196-9

Exercise Workbook for Advanced AutoCAD 2002, ISBN 0-8311-3197-7

Exercise Workbook for Beginning AutoCAD 2000, 2000i, and LT, ISBN 0-8311-3194-2

Exercise Workbook for Advanced AutoCAD 2000, 2000i, and LT, ISBN 0-8311-3195-0

Exercise Workbook for Advanced AutoCAD 2000, ISBN 0-8311-3193-4

All books can be ordered from
Industrial Press
Toll Free: 888-528-7852   or on line at   www.industrialpress.com

EXERCISE WORKBOOK

for

Advanced AutoCAD®

2004

by
Cheryl R. Shrock

Chairperson
Drafting Technology
Orange Coast College, Costa Mesa, Ca.

Registered Author/Publisher

Original Edition 2004

Published by:
Shrock Publishing
P.O. Box 9767
Newport Beach, Ca. 92658
support@shrockpublishing.com
714 557-2555

Limits of Liability and disclaimer of Warranty

The author and publisher make no warranty of any kind, expressed or implied, with regard to the documentation contained in this book.

AutoCAD $^®$ 2004 is a registered trademark of Autodesk, Inc.

ISBN 0-8311-3199-3

Visit our website: www.industrialpress.com

Table of Contents

Lesson 20

PROJECTS

Architectural
Electro-mechanical
Mechanical

APPENDICES

Appendix A. Add a Printer / Plotter
Appendix B. Printing in Color or Black <u>with Lineweights</u>
Appendix C. How to Create a New "Color-dependent Plot Style" to be used
             <u>without Lineweights</u>

INDEX

INTRODUCTION

About this workbook
This workbook is designed to <u>follow</u> *Exercise Workbook for <u>Beginning</u> AutoCAD 2004*. It is excellent for classroom instruction or self-study. There are 20 lessons and 3 *on-the-job* type projects in Architectural, Electro-mechanical and Mechanical.

Lessons 1 through 12, combined with lessons 1 through 30 in the <u>Beginning</u> workbook, complete the basic 2D commands.

Lessons 13 through 20 introduce you to many basic 3D commands.

Each lesson starts with step-by-step instructions followed by exercises designed for practicing the commands you learned within that lesson. The *on-the-job* projects are designed to give you more practice in your desired field of drafting.

Exercises that include printing are designed for a Hewlett Packard 4MV LaserJet printer (11 x 17 sheet size) and a Hewlett Packard 500 or 600 Design Jet (18 x 24 & 24 x 36 sheet size). You should configure your system to recognize their specifications in order to *plot preview* the exercises. *Note: your computer does not have to be attached to a printer to configure it. Only the printer specifications and limitations will be loaded.* Refer to **Appendix A "Add a Printer".**

IMPORTANT:
Two files are required. "1Workbook Helper.dwt and "3D Helper.dwg". Refer to page Intro-2 for instructions on "downloading" and "creating templates".

About the Author
Cheryl R. Shrock is a Professor and Chairperson of Computer Aided Design at Orange Coast College in Costa Mesa, California. She is also an Autodesk® registered author / publisher. Cheryl began teaching CAD in 1990. Previous to teaching, she owned and operated a commercial product and machine design business where designs were created and documented using CAD. This workbook is a combination of her teaching skills and her industry experience.

"Sharing my industry and CAD knowledge has been the most rewarding experience of my career. Students come to learn CAD in order to find employment or to upgrade their skills. Seeing them actually achieve their goals, and knowing I helped, is a real pleasure. If you read the lessons and do the exercises, I promise, you will not fail."

Cheryl R. Shrock

CREATE A TEMPLATE

Note: If you have already created the template "1Workbook Helper.dwt", required in the "2004 Beginning Workbook", skip to Lesson 1.

The first item on the learning agenda is how to create a template file from the **"1Workbook Helper.dwg"**. Go to the website **www.shrockpublishing.com** and download the files for workbooks 2004 and save them to a disk.

Now we will create a template. This will be an easy task.

1. Start AutoCAD:
    Start button / Programs / Autodesk / AutoCAD 2004 or LT / AutoCAD 2004 or LT

    Note: If a dialog box appears select the "Cancel" button.

2. Select **File / Open**

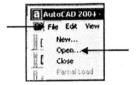

3. Select the **Directory** in which the downloaded files were saved. (Click on the ▼)

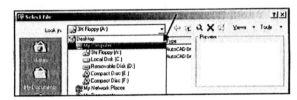

4. Select the file "**1Workbook Helper.dwg**" and then "**Open**" button.

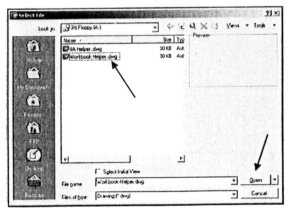

Notice the 3 letter extension for a "drawing" file is ".dwg"

5. Select **"File / Save As..."**

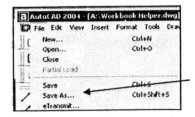

6. Select the "**Files of type:**" down arrow ▼ to display different saving formats. Select "**AutoCAD Drawing Template (*.dwt)**"

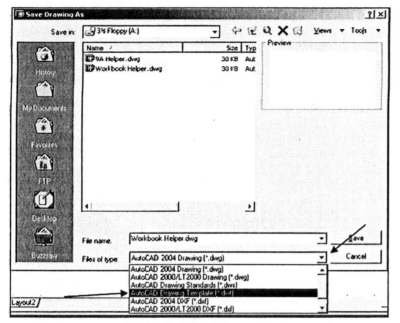

Notice the 3 letter extension for Template is ".dwt"

A list of all the AutoCAD templates will appear.

7. Type the new name "**1Workbook Helper**" in the "**File name:**" box and then select the "**Save**" button.

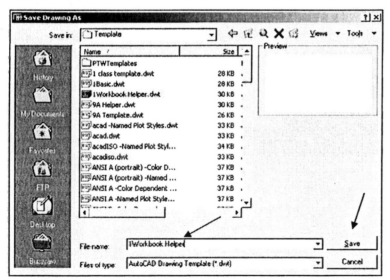

The "1" in the name places the file at the top of the list. AutoCAD lists numbers first and then alphabetical.

Notice it was not necessary to type the extension .dwt because "Files of type" was previously selected.

8. Type a description and the select the "OK" button.

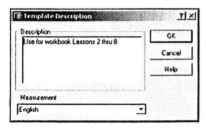

Now you have a template to use for lessons 2 through 8. At the beginning of each of the exercises you will be instructed to open this template. Using a template as a master setup drawing is good CAD management.

OPENING A TEMPLATE

The template that you created on the previous page will be used for lessons 1 and 2 only. It will appear as a blank screen but there are many variables that have been preset. This will allow you to start drawing immediately. (*If you have not completed the exercises in "Exercise Workbook for Beginning AutoCAD 2004", I strongly suggest that you do. It will make learning AutoCAD less confusing and more fun.*)

Let's start by opening the "1Workbook Helper.dwt" template.

1. Select **FILE / NEW**.

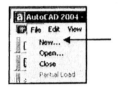

2. Select the **Use a Template** box. (third from the left)

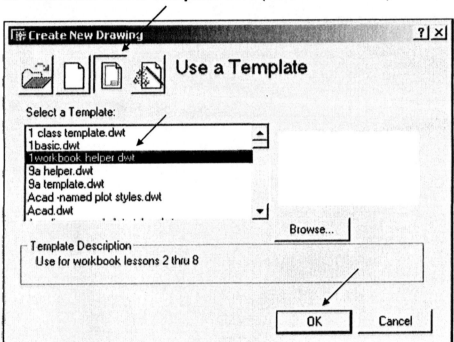

3. Select **1workbook helper.dwt** from the list of templates.

   (NOTE: If you do not have this template, refer to page Intro-2)

4. Select the **OK** button

NOTE: *If you find that you have more than one drawing open you have not configured your AutoCAD software to allow only one drawing open at one time. In Lesson 1 you will configure AutoCAD for multiple open drawings. Refer to Intro-7 for "Single drawing compatibility mode" setting under "General Options". Check this option box.*

Configuring your system

While you are using this workbook, it is necessary for you to make some simple changes, to your configuration, so our configurations are the same. This will ensure that the commands and exercises work as expected. The following instructions will walk you through those changes. *Note: If you have already configured your system for the 2004 "Beginning Workbook" you may skip to Lesson 1.*

A. First start AutoCAD® 2004.
　　1. Click "Start" button in the lower left corner of the screen.
　　2. Choose "Programs / Autodesk / AutoCAD 2004 or LT / AutoCAD 2004 or LT
　　3. You should see a blank screen. (If the "Create a new drawing" dialog box appears, select "Cancel" and continue.

B. At the bottom of the screen there is a white rectangular area called the "*Command Line*". Type: **_Options_** then press the **<enter>** key. (not case sensitive)

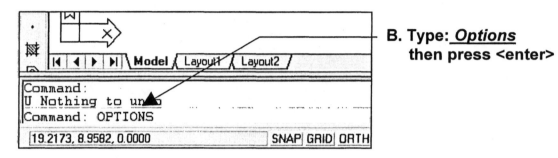

B. Type: _Options_
　　then press <enter>

C. Select the **_Display_** tab and change the settings on your screen to match the dialog box below. Pay special attention to the settings with an ellipse around them.

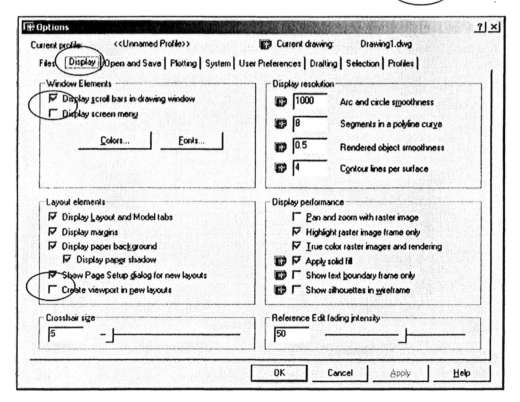

E. Select the **Open and Save** tab and change the settings on your screen to match the dialog box below.

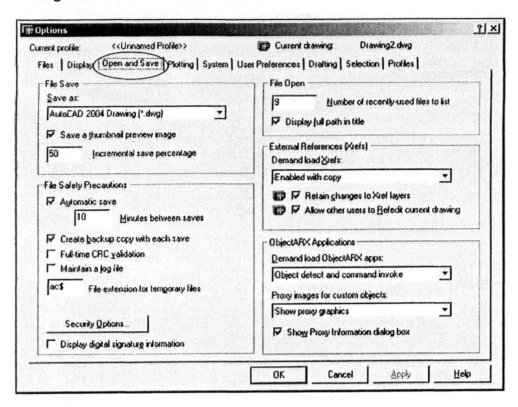

F. Select the **Plotting** tab and change the settings on your screen to match the dialog box below.

IMPORTANT: Add this printer.
See Appendix A for instructions
(Don't worry, it is not difficult)

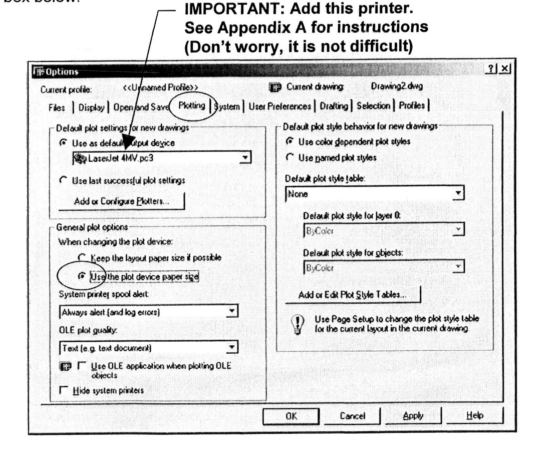

G. Select the *System* tab and change the settings on your screen to match the dialog box below.

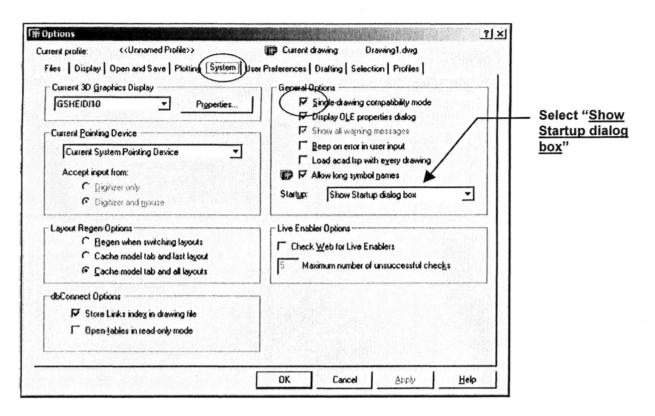

Select "**Show Startup dialog box**"

H. Select the *User Preferences* tab and change the settings on your screen to match the dialog box below.

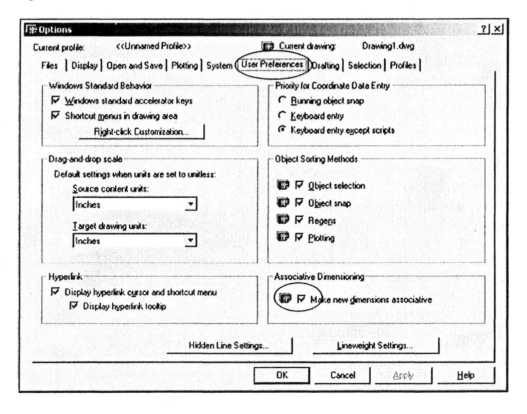

I. Select the **Right-click Customization..** box and change the settings on your screen to match the dialog box below.

Select "Right-click customization" button

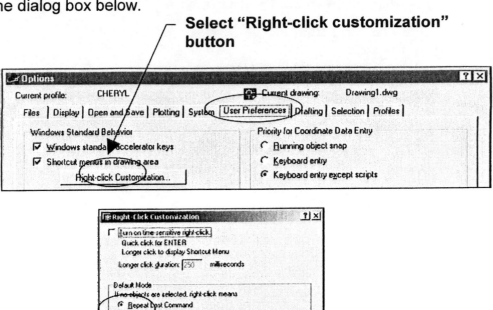

J. Select the **Apply & Close** button, shown above, before going on to the next tab.

K. Select the **Drafting** tab and change the settings on your screen to match the dialog box below.

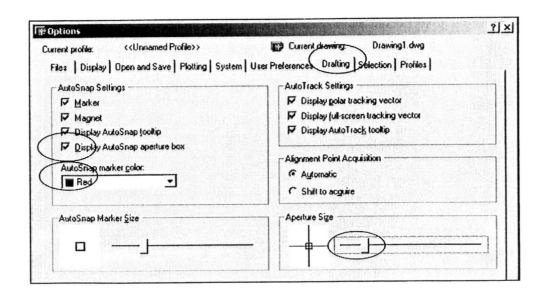

L. Select the *Selection* tab and change the settings on your screen to match the dialog box below.

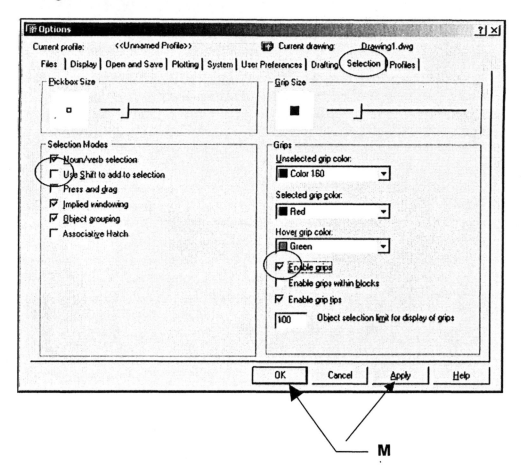

M. Select the *Apply* button then the *OK* button.

Customizing your Wheel Mouse

A Wheel mouse has two or more buttons and a small wheel between the two top side buttons. The default functions for the two top buttons are as follows:
Left Hand button is for **input.**
Right Hand button is for **Enter** or the **shortcut menu**.

You will learn more about this later. But for now follow the instructions below.

Using a Wheel Mouse with AutoCAD®

To get the most out of your Wheel Mouse set the **MBUTTONPAN** setting to **"0"** as follows:

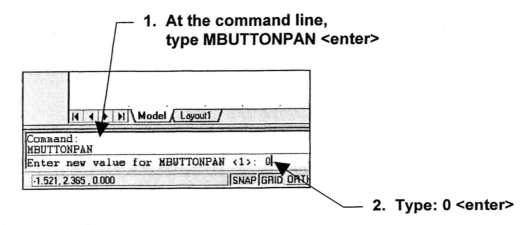

1. At the command line, type MBUTTONPAN <enter>

2. Type: 0 <enter>

After you understand the function of the "Mbuttonpan" variable, you can decide whether you prefer the setting "0" or "1" as described below.

MBUTTONPAN setting 0:

ZOOM   Rotate the wheel forward to zoom in
            Rotate the wheel backward to zoom out

OBJECT   Object Snap menu will appear when you press the wheel
SNAP

MBUTTONPAN setting 1:

ZOOM   Rotate the wheel forward to zoom in
            Rotate the wheel backward to zoom out

ZOOM
EXTENTS   Double click the wheel

PAN   Press the wheel and drag

LEARNING OBJECTIVES

Note:
This lesson should be used to determine whether or not you are ready for this level of instruction. If you have difficulty creating Exercises 1A, B and C, you should consider starting with the "Exercise Workbook for Beginning AutoCAD 2004".

LESSON 1

DISPLAY MULTIPLE DRAWINGS

AutoCAD has an option that allows you to have one or more drawings open at the same time. This option is **"Single-drawing compatibility mode".** If you have "Single-drawing compatibility mode" **ON**, only one drawing will be open on the screen. If you have "Single-drawing compatibility mode" **OFF**, multiple drawings may be open on the screen.

You may view the multiple drawings side by side (tiled) or as a full screen (cascade).
See examples of each on the next page.

How to configure AutoCAD to open *"Multiple drawings".*

1. Select **TOOLS / OPTIONS**
2. Select the **SYSTEM** tab.
3. Remove the **check mark** from "Single-drawing compatibility mode" box.
4. Select **APPLY** and the **OK** button.

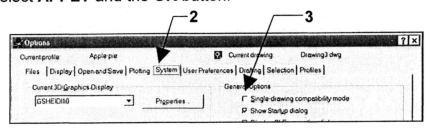

How to CLOSE multiple drawings.

1. Select the **Window** pulldown menu.
2. Select **Close** to close one drawing or **Close All** to close all the drawings at once.

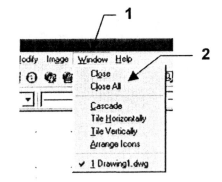

How to view multiple drawings.

1. Select the **WINDOW** pulldown menu.
2. Select Cascade, Tile Horizontally or Tile Vertically.

(Examples on next page)

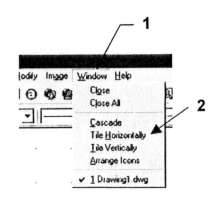

Example of "TILED"

Tiled <u>Horizontally</u>

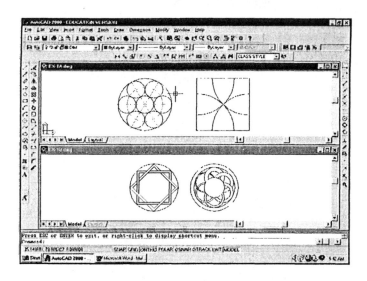

Tiled <u>Vertically</u>

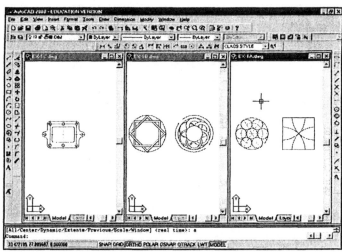

Example of "CASCADE".

To view one of the multiple drawings use one of the following:

1. Use **Crtl + tab** to toggle between drawings.
     or
2. Select **Window** pulldown menu, then select the drawing name from the list shown
     or
3. Click on the **Title bar**

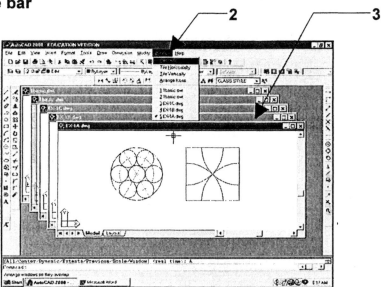

WARM UP DRAWINGS

The following drawings have been included for two purposes:

First purpose: To make sure you remember the commands taught in the Beginning Workbook and to get you prepared for the new commands in this Advanced workbook.

Second purpose: To confirm that you are ready for this level of instruction.

IMPORTANT, PLEASE READ THE FOLLOWING:

*This workbook assumes you already have enough basic AutoCAD 2004 knowledge to easily complete Exercises 1A, 1B and 1C. If you have difficulty with these exercises, you should consider reviewing "**Exercise Workbook for Beginning AutoCAD 2004**". If you try to continue, without this knowledge, you will probably get confused and frustrated. It is better to have a good solid understanding of the basics before going on.*

NOTE:
The template file mentioned through out the workbook is the same template that was used in the Beginning Workbook. If you do not have the template refer to page Intro-2 for instructions.

Note to Instructors

The Page Set ups in this workbook are designed for HP plotters, 4MV, 500 AND 600. If your plotters are different, you will need to revise the plotting instructions throughout the lessons.

EXERCISE 1A

INSTRUCTIONS:
1. Start a **NEW** file and select **1workbook helper.dwt**.
2. Draw the Objects below using Layers: Object and Center.
3. Do not Dimension.
4. Add your name anywhere on the drawing. Use Layer: Txt-Lit Style: Standard Height: .25
5. Save as: EX-1A
6. Plot from **"Model"** using the instructions on page 1-8.

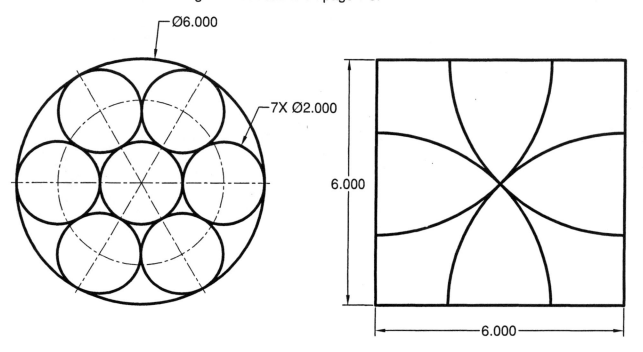

Ø6.000

7X Ø2.000

6.000

6.000

CONSTRUCTION HINTS

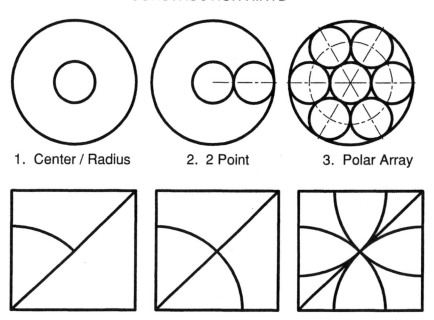

1. Center / Radius   2. 2 Point   3. Polar Array

1. Arc: Center, Start, End   2. Extend   3. Mirror or Polar Array

EXERCISE 1B

INSTRUCTIONS:
1. Start a **NEW** file and select **1workbook helper.dwt**.
2. Draw the Objects below using Layer: Object.
3. Do not Dimension.
4. Add your name anywhere on the drawing. Use Layer: Txt-Lit Style: Standard Height: .25
5. Save as: EX-1B.
6. Plot from **"Model"** using the instructions on page 1-8.

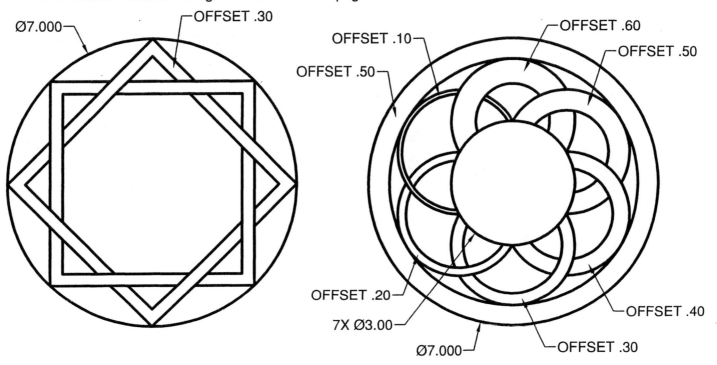

CONSTRUCTION HINTS

1. 8 Sided Polygon   2. 4 Sided Polygons   3. Offset and Trim

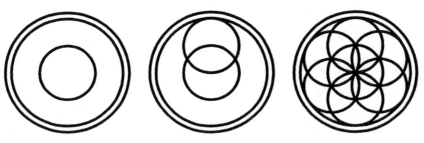

1. Circle, Cen / Rad   2. Circle, 2P   3. Array, Offset, Trim

EXERCISE 1C

INSTRUCTIONS:
1. Start a **NEW** file and select **1workbook helper.dwt**.
2. Draw the Objects below using Layers: Object, Center and Dimension.
3. Dimension using Dimension Style **"Class Style"**.
4. Add your name anywhere on the drawing. Use Layer: Txt-Lit Style: Standard Height: .25
5. Save as: EX-1C
5. Plot from **"Model"** using the instructions on page 1-8.

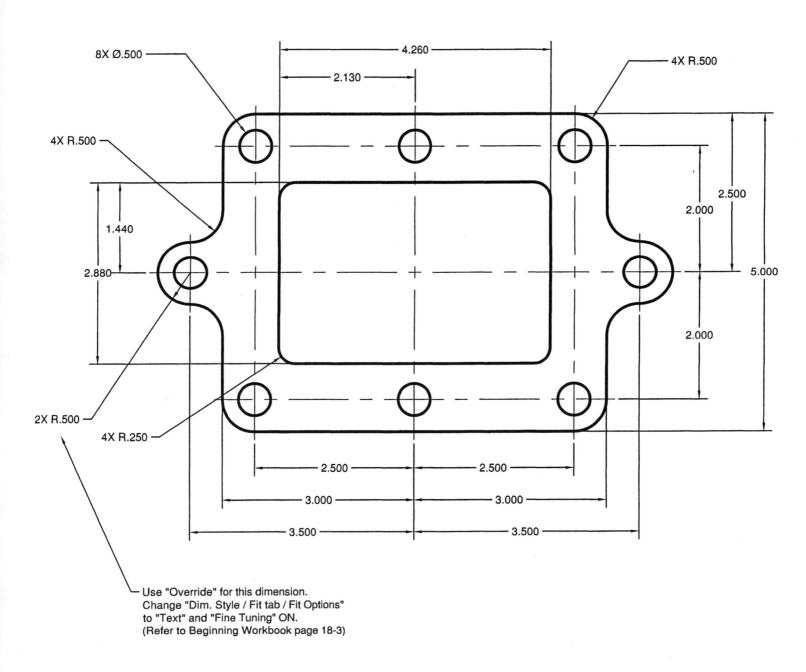

8X Ø.500

4X R.500

4X R.500

1.440

2.880

2X R.500

4X R.250

4.260

2.130

4X R.500

2.500
2.000

5.000

2.000

2.500

2.500

3.000

3.000

3.500

3.500

Use "Override" for this dimension.
Change "Dim. Style / Fit tab / Fit Options"
to "Text" and "Fine Tuning" ON.
(Refer to Beginning Workbook page 18-3)

REVIEW PLOTTING FROM MODEL SPACE

1. **Important:** Open the drawing you want to plot.
2. Make sure that the Model tab is selected.   (Model tab)
3. Select the **Plot** command by "right clicking" on the "Model" tab or using one of the following methods listed below:

> **Type = Print or Plot**
> **Pulldown = File / Plot**
> **Tool bar = Standard**

The Plot dialog box below should appear.

4. Select the **"Plot Device"** tab.

5. Select the Printer (Plot device)
   Note: This printer represents a size 17 x 11. If it is not in the list, it needs to be configured.
   Refer to Appendix A

   Notice: You may configure a printer even though your computer is not attached to it.

6. Select the Plot Style Table Select "**All Black.ctb**".

   (If this .ctb file is not in the list, refer to "Exercise Workbook for Beginning AutoCAD 2004, page 9-8.)

7. Select the "**Plot Settings**" tab.

8. Check all the settings to make sure they match the workbook example.

9. Set the scale to 1:1

10. Select **Full Preview** button.

11. If your drawing looks correct, press the **ESC** key then **OK**.

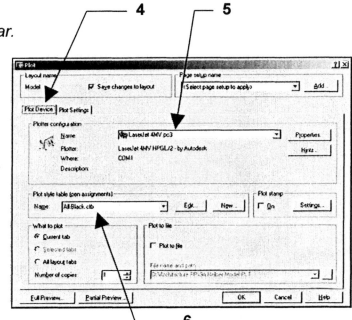

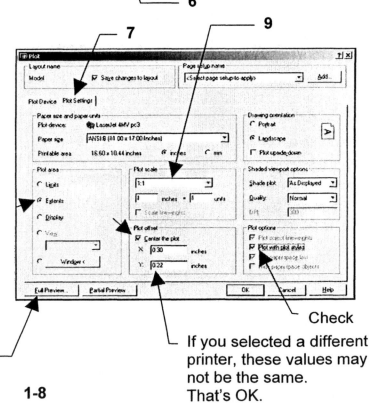

Check

If you selected a different printer, these values may not be the same. That's OK.

LEARNING OBJECTIVES

After completing this lesson, you will be able to:

1. Modify or Rename a toolbar.
2. Create a new toolbar.
3. Restore default toolbars
4. Customize the Status Bar
5. Redo multiple commands
6. Understand Profiles
7. Create a new Profile
8. Import and use a Profile

LESSON 2

HOW TO CUSTOMIZE A TOOLBAR

You probably have noticed that you are using some toolbar buttons more than others. Some you don't use at all. AutoCAD allows you to customize your toolbars for quick access. You can delete infrequently used buttons, add buttons or create an entirely new toolbar and fill it with buttons of your choice. It is not difficult.

HOW TO MODIFY A TOOLBAR

1. Select **View / Toolbars...**
2. Make sure the toolbar you want to modify is open.

Note: You can only modify a toolbar when the "Customize" dialog box is <u>displayed</u>.
You can modify any <u>displayed</u> toolbar as follows:

Reposition a button on a toolbar
Place the cursor on the button, press and hold down the left mouse button. Now slide the cursor to the new location on the toolbar. (You will see a black vertical bar as you move the cursor) When the black bar is in the correct location, release the mouse button.
(This is called "**Click and drag**")

Remove a button from a toolbar
"Click and drag" the button off the toolbar then release mouse button.

Copy a button from one toolbar to another
Hold the Ctrl key down while dragging the desired button from one toolbar to another.

Add a button to a toolbar
1. Select the "Commands" tab.
2. Select the toolbar where desired button is located.
3. Drag the selected button to the toolbar.

HOW TO RENAME A TOOLBAR

1. Select the **Toolbars** tab.
2. Highlight the toolbar you want to rename.
3. Select the **Rename** button.
4. Type the new name, then OK.

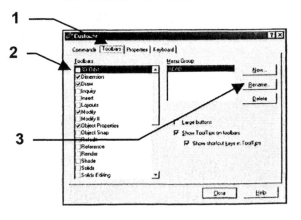

HOW TO CREATE A NEW TOOLBAR

You first have to create an empty toolbar and then drag buttons onto the new toolbar.

1. Select the **Toolbars** tab.
2. Select the **New** button.

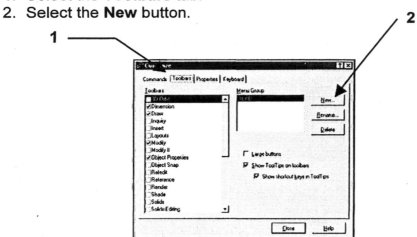

3. Type the new name, then OK.

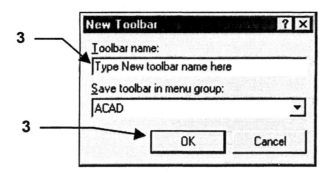

Note: Sometimes it is difficult to find the new toolbar on the screen because it is very small. Look at the top middle of the screen, it is usually there.

Your new Toolbar will
look like this.

4. Refer to the **MODIFY** instructions, on page 2-2, to add buttons to the new toolbar.

HOW TO RESTORE THE DEFAULT TOOLBARS

After you have experimented with customizing toolbars you may wish to restore the default toolbars. The following instructions will guide you through the process.

1. At the command line type: **MENU <enter>**
2. Select **"menu template (*.mnu)"** from the "File of type" list. (See below)

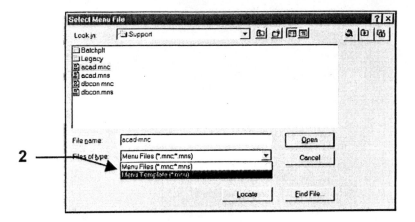

3. Select the **acad.mnu** file then the **OPEN** button.

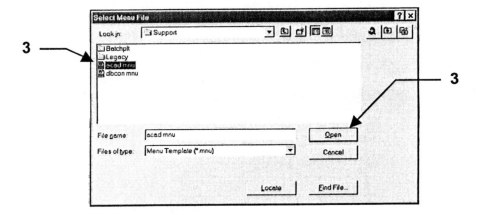

4. The following warning will appear. Read it and be sure that you want to lose all of your customized toolbars and menus. If this is OK, select **YES**. If you do not want to lose your customized toolbars and menus, select **NO.**

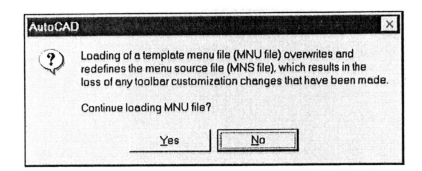

HOW TO CUSTOMIZE THE STATUS BAR

AutoCAD allows you to customize the Status Bar, located at the bottom of the AutoCAD window. For example, some people never use "OTRACK", so they could remove the button. *Personally, I see no reason to remove any buttons but you can make that decision.*

1. Click the small down arrow at the right end of the status bar.

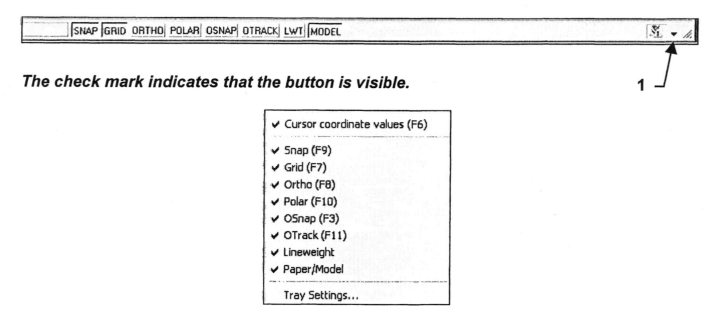

The check mark indicates that the button is visible.

2. Click on the button name and the check mark will disappear and the button will disappear from the Status Bar.

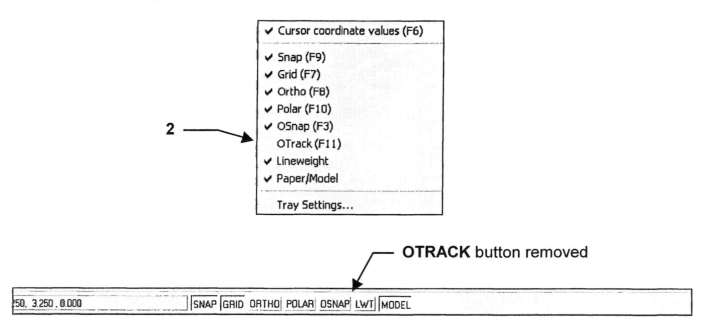

Note:
To replace the button, click on the button name again, following steps 1 and 2 above.

REDO multiple commands

AutoCAD has created a new MREDO command to redo multiple UNDO commands.
You can redo recently undone commands from the Redo drop-down list.
Remember, <u>redoing a command</u> only works if you have <u>Undone a command</u>.

Note:
If the undo or redo arrows are grayed out, it means that there are no objects drawn and/or no undo commands have been issued.

How to redo multiple undo commands.

1. Start a new drawing.
2. Draw a line, rectangle and a circle.
3. Copy all 3 objects.

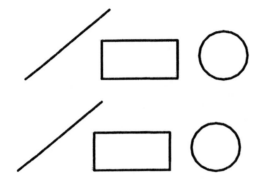

Your drawing should look approximately like this.

4. Click on the UNDO drop-down arrow.

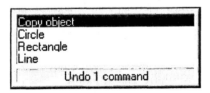

The UNDO list displays the creating and editing commands with the most recent at the top of the list.

5. Highlight the Copy object, Circle and Rectangle command and press the left mouse button, leaving only the Line command.

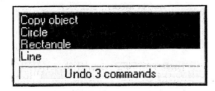

The Rectangles and Circles should have disappeared and only the Lines remain.

Now let's "Redo" some of what we "Undo"ed.

6. Click on the REDO drop-down arrow.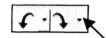

7. Highlight the Rectangle and Circle commands and press the left mouse button.

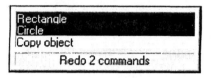

One Rectangle and one Circle should have reappeared.

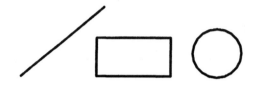

Understanding USER PROFILES

What is a User Profile?

Some computers are shared by multiple users; such as students in a classroom or employees at work. Each of these users have individual preferences on how they want the AutoCAD screen to appear. For example, you may prefer a white background instead of a black background color. You may also prefer a different length for the cursor crosshairs and placement of toolbars. All of these changes, and more, are considered the user's **profile**.

In the following exercises, you will learn how to:

1. Create a profile.
2. Save (export) the profile for future use.
3. Retrieve (import) the profile.

EXERCISE 2A

CREATE A NEW PROFILE.

Although there are many settings and values you can change, we will limit the changes to the following:

1. AutoCAD screen colors
2. Cross hair size
3. AutoSnap marker size
4. Aperture size
5. Pick box size
6. Enable grips
7. Grip size
8. Tool bar preferences

Note: If you are in a classroom, you should ask the instructor for approval before changing any settings other than the 8 listed above.

A. Select **Tools / Options...** or move the cursor to the command line and press the right mouse button then select **"OPTIONS"** from the short cut menu.

B. Select the **Profiles** tab.
   1. Select the **"Add to List"** button.

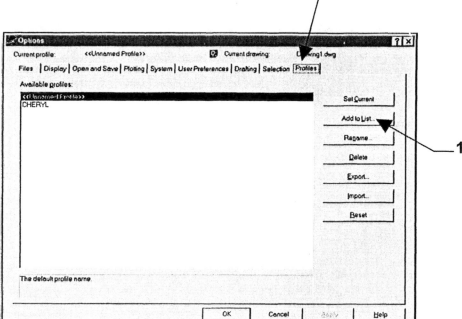

2. Type the profile name. *(Type your name: Last-First) but* **DO NOT** press <enter>
   Press the **TAB** key.

3. Type a description, such as: *This is the profile (your name here) uses in class.*

4. Select the **"Apply and Close"** button.

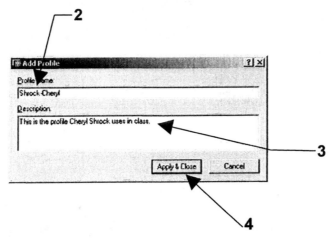

5. Select the Profile name you just created. (Click on it)

6. Select the **"Set Current"** button.

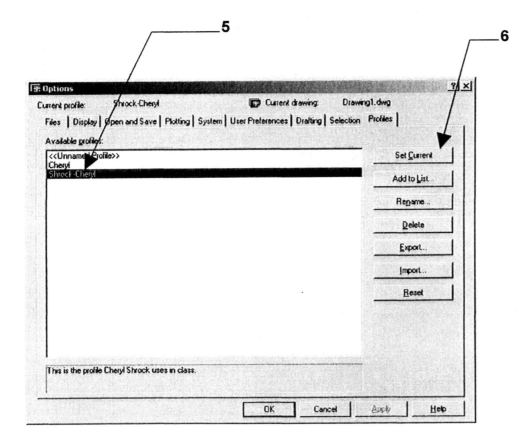

C. Select the **DISPLAY** tab.

1. Change the **"Crosshair Size"** to your preference from 5 to 100.
   The number represents a percentage of the screen size.
   5 = 5% of the screen (default)   100 = 100% of the screen size

2. Select the **Colors...** button.

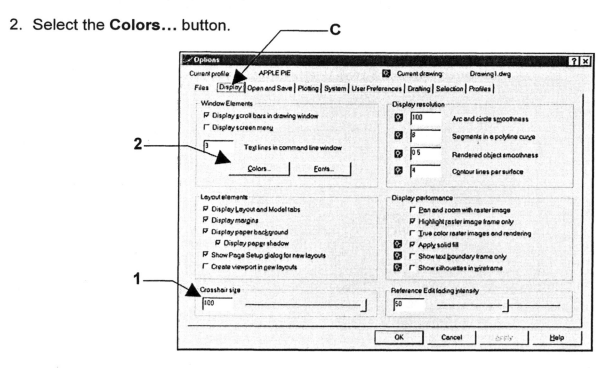

YOU CAN CHANGE THE COLORS OF:
   Model and Layout: background and pointer
   Command Line: background and text
   Auto Track Vector Color

3. Click on the area you would like to change or select by name.

4. Select the color for that area.

5. When complete, select the **"Apply & Close"** button.

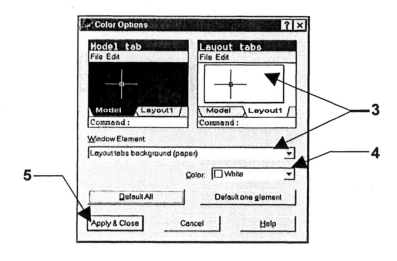

D. Select the **DRAFTING** tab.

1. Adjust the size of the "**AutoSnap Marker**" to your preference. Click and Drag the Slider Button to the left (min) or right (max). This is the marker that appears on objects at the preset object snap locations when using AutoSnap.

2. Adjust the size of the "**Aperture**" to your preference. Click and Drag the Slider Button to the left (min) or right (max). The aperture box appears when using **AutoTRACKING**.

3. When complete, select the **Apply Button**.

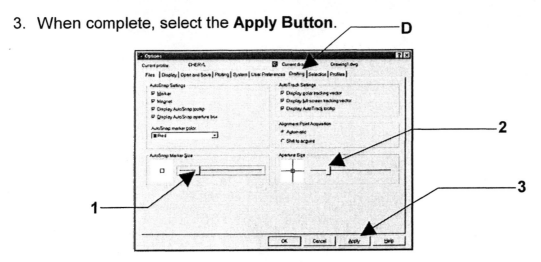

E. Select the **SELECTION** tab.

1. Adjust the **Pickbox Size** to your preference. The pickbox appears when *selecting objects. (The size will affect the selection precision)*

2. Enable (on) the display of grip boxes by placing a check mark in the box. Disable (off) by removing the check mark.

3. Adjust the size of the **Grip** boxes.

4. When complete, select the **Apply Button.**

5. Select the **OK** button to leave the **Options** dialog box.

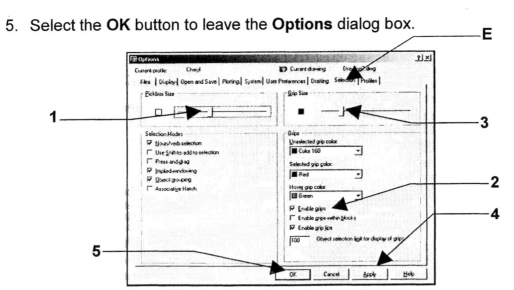

F. TOOLBARS

Open preferred toolbars and close the unwanted toolbars.
Move the open toolbars to your desired location.

Now look at the screen. Is everything to your preference? If so, we will now save this Profile to a disk. This will allow you to transfer this profile to other computers.
(Note: the profile will automatically be saved to the computer you are working on right now.)

G. SAVING A PROFILE.

1. Select **Tools / Options...** or move the cursor to the command line and press the right mouse button then select **"OPTIONS"** from the short cut menu.

2. Select the **PROFILES** tab.

3. Select the **EXPORT** button.

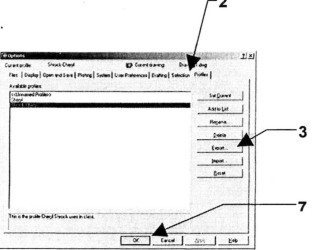

4. Select a "removable" disk, such as a 3-1/2 floppy or Zip disk.

5. Type the name of the profile. *(Note: the file extension is **.ARG** for profiles)*

6. Select the **SAVE** button.

7. Now select the **OK** button.
   (See dialog box above)

Note: If you make changes to your profile you must Export (save) the profile again as explained above.

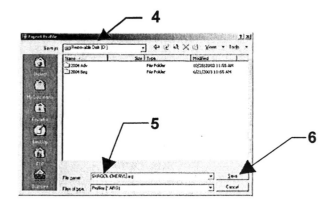

If you want to move your profile to another computer, you must IMPORT. (Refer to page 2-14)

EXERCISE 2B

IMPORTING A USER PROFILE.

Now that you have learned how to created a profile, you will want to use it.

If the profile "does" exist on the computer, you merely select it and then select **"SET CURRENT"**.

If the profile "does not" already exist on the computer you are working on, you must **IMPORT** it and then "**SET CURRENT**".   (Follow the instructions below)

A. Select **Tools / Options...** or move the cursor to the command line and press the right mouse button then select **"OPTIONS"** from the short cut menu.

B. Select the **PROFILES** tab.
   1. Select the **IMPORT** button.

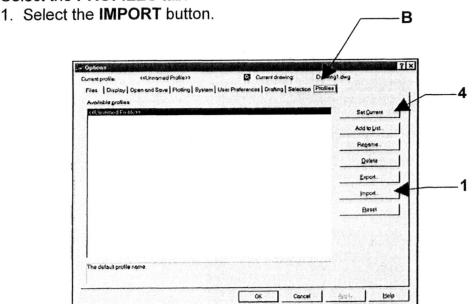

2. Find and select your *profile* file

3. Select the **OPEN** button.

4. Select "**Set Current**"

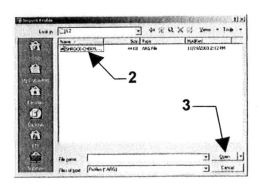

LEARNING OBJECTIVES

After completing this lesson, you will be able to:

1. Create a Setup drawing.
2. Create Borders for 17 x 11, 24 x 18 and 36 x 24
3. Lock a Viewport
4. Create Page Set ups for 17 x 11, 24 x 18 and 36 x 24.

LESSON 3

EXERCISE 3A
CREATE A MASTER DECIMAL SETUP DRAWING

The following instructions will guide you through creating a "Master" decimal setup drawing. The "1Workbook Helper" is an example of a Master setup drawing. Even though the screen appears blank, the actual file is full of settings, such as: Units, Drawing Limits, Snap and Grid settings, Layers, Text styles and Dimension Styles. Once you have created this "Master" drawing, you just open it and draw. No more repetitive inputting of settings.

NEW SETTINGS

A. Begin your drawing without a template as follows:

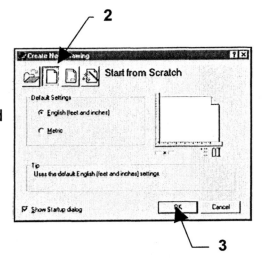

   1. Select **"FILE / NEW"**
   2. Select **"START FROM SCRATCH"** Box.
   3. Select **"OK"**.
   4. Your screen should be blank, no grids and the current layer is 0

B. Set drawing specifications as follows:

   1. Set **"UNITS"** of measurement
     Use "**FORMAT / UNITS** and change the settings as shown then select **OK.**

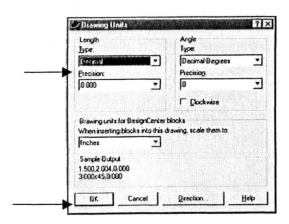

   2. Set **"DRAWING LIMITS"** (Size of drawing area)
     USE "**FORMAT / DRAWING LIMITS**
        a. Lower left corner = 0.000,0.000
        b. Upper right corner = 17 , 11
        c. Use **"VIEW / ZOOM / ALL"** to generate the new limits
        d. Set your **Grids** to **ON** to display the paper size.

3. Set **"SNAP AND GRID"**
   Use **"TOOLS / DRAFTING SETTINGS**

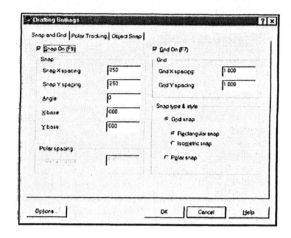

4. Set **"PICKBOX"** size (your preference)
   See page 2-12 E-1 for instructions.

NEW LAYERS

C. Create new layers

1. First the **Load** linetypes listed below. (Use Format / Linetype / Load)
   CENTER2
   HIDDEN
   PHANTOM2

2. Assign names, colors, linetypes and plotability. (Use Format / Layer)

NAME	COLOR	LINETYPE	LWT	PLOT
BORDER	RED	CONTINUOUS	.039	YES
CENTER	CYAN	CENTER2	Default	YES
CONSTRUCTION	WHITE	CONTINUOUS	Default	NO
DIMENSION	BLUE	CONTINUOUS	Default	YES
HATCH	GREEN	CONTINUOUS	Default	YES
HIDDEN	MAGENTA	HIDDEN	Default	YES
OBJECT	RED	CONTINUOUS	.024	YES
PHANTOM	MAGENTA	PHANTOM2	Default	YES
SECTION	WHITE	PHANTOM2	.031	YES
SYMBOL	RED	CONTINUOUS	.024	YES
TEXT HEAVY	WHITE	CONTINUOUS	Default	YES
TEXT LIGHT	BLUE	CONTINUOUS	Default	YES
THREADS	GREEN	CONTINUOUS	Default	YES
VIEWPORT	GREEN	CONTINUOUS	Default	NO
XREF	WHITE	CONTINUOUS	Default	YES

NEW TEXT STYLE

D. Create a text style

   1. Select "**FORMAT / TEXT STYLE**"
   2. Create a NEW text style named Class Text.
   3. Select the Font and Effects shown in the dialog box below.

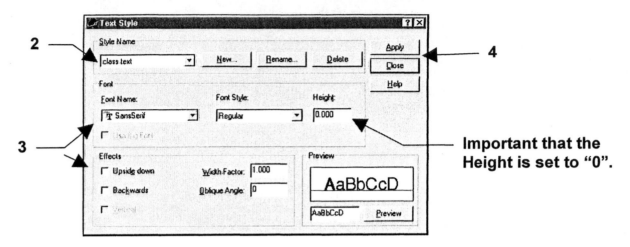

   4. When complete, select **APPLY** then **CLOSE**.

NEW DIMENSION STYLE

E. Create a new Dimension Style named *Class Style*

   1. Select the **DIMENSION / STYLE** command
   2. Select the **NEW** button.
   3. Make the changes to the following dialog boxes.

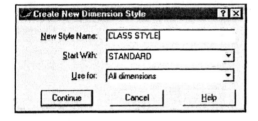

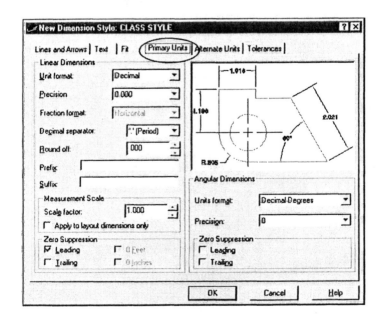

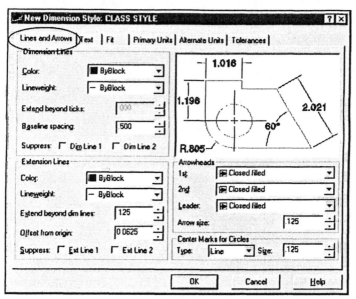

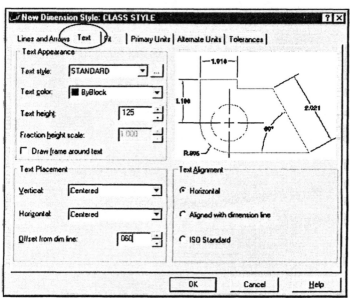

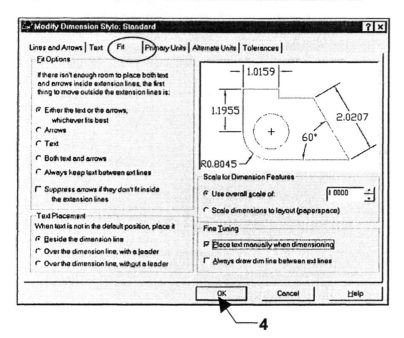

4. After changing the settings shown above, select the **OK** button.

5. *Your new style "Class Style" should be listed.* Select the **"Set Current"** button to make your new style "Class Style" the style that you will use.

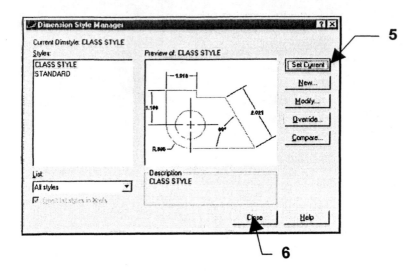

6. Select the **Close** button.

THIS NEXT STEP IS VERY IMPORTANT..

F. SAVE ALL THE SETTINGS YOU JUST CREATED
1. Select File / Save as
2. Save as: **My Decimal Setup**

NOTE: This was just step 1 to creating your setup drawing. Continue on through the following exercises, in this lesson, using this same file (My Decimal Setup). Each step adds more information to this drawing.

EXERCISE 3B
CREATE A BORDER FOR PLOTTING

The following instructions will guide you through creating a Border drawing that will be used in combination with "My Decimal Setup" when plotting. You will create a Layout and draw a border with a title block. All of this information will be saved and you will not have to do this again.

A. Open **My Decimal Setup**

B. Select a **LAYOUT1** tab.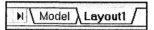

Note: If the Page Setup dialog box shown below does not appear automatically, right click on the Layout tab, then select Page Setup.

1. Type the new name:
   11x17 (1 to 1)

2. Select the "*Plot Device*" tab.

3. Select the Plotter.
(Use this plotter for this exercise.
Refer to Appendix A for instruction
if it does not appear in the list.)

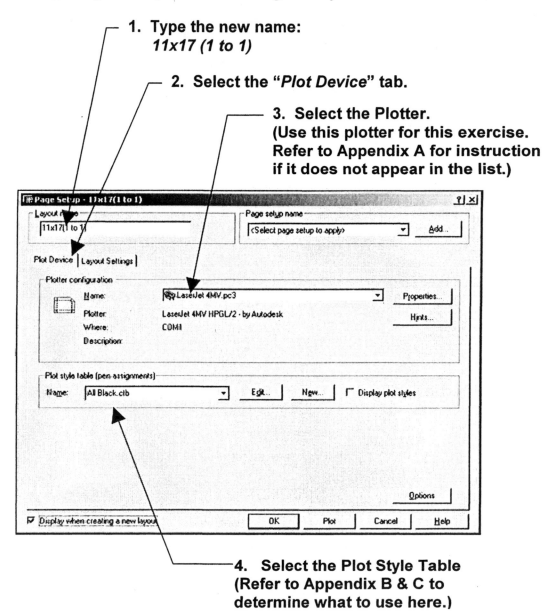

4. Select the Plot Style Table
(Refer to Appendix B & C to
determine what to use here.)

5. Select the "*Layout Settings*" tab.

6. Select the "*Paper Size*".

7. Select scale "1:1"

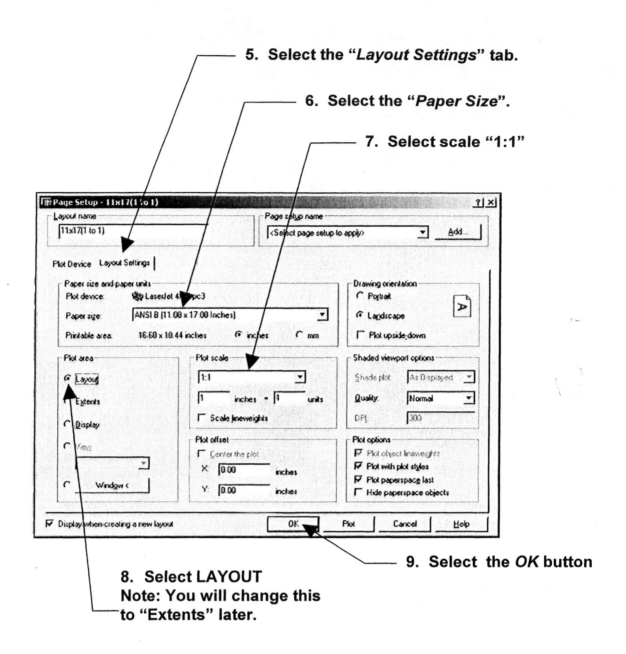

8. Select LAYOUT
Note: You will change this
to "Extents" later.

9. Select the *OK* button

You should now have a sheet of paper displayed on the screen and the Layout tab should now be displayed as "11x17 (1 to 1)".

This sheet is in front of "Model". In Exercise 3C, you will cut a hole (viewport) in this sheet so you can see through to Model.

C. Draw the rectangular border with title block, shown below, on the sheet of paper shown on the screen. (Don't forget the @ symbol; @16, 10)

D. When you have completed the Border, shown below:
1. Select File / Save as
2. Save as: **My Decimal Setup** (Again)

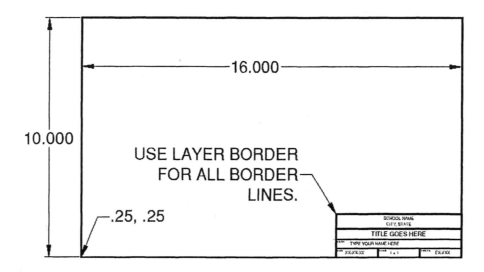

Important: Use "Single Line Text" to place the text in the title block.

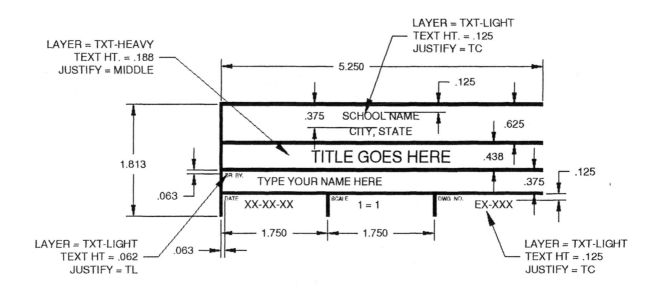

Don't forget to save, refer to "D" above.

EXERCISE 3C
CREATE A VIEWPORT

The following instructions will guide you through creating a VIEWPORT in the Border Layout sheet. Creating a viewport has the same effect as cutting a hole in the sheet of paper. You will be able to see through the viewport frame (hole) to Model.

A. Open **My Decimal Setup**

B. Select the **11-17 (1 to 1)** tab.

C. Select layer "Viewport"

D. Select the Single Viewport icon from the Viewports Toolbar or Type MV <enter>.

Single Viewport icon (LT's toolbar looks a little different but it works the same)

E. Draw a Single viewport approximately as shown.

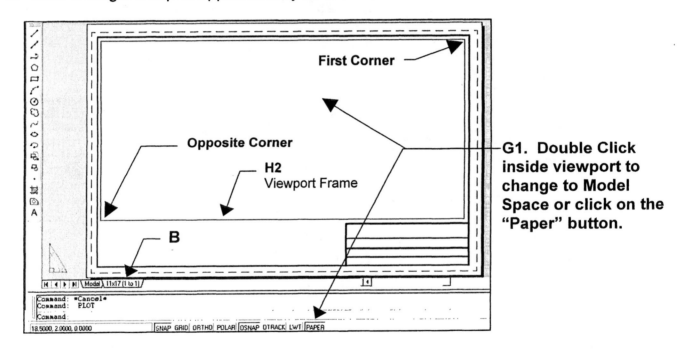

First Corner

Opposite Corner

H2 Viewport Frame

B

G1. Double Click inside viewport to change to Model Space or click on the "Paper" button.

F. After successfully creating the Viewport, you should now be able to see through to Model. (Your grids should appear if they are ON.)

G. Adjust the Model space scale.
   1. Select the **Paper** button so it changes to "Model" or double click inside the Viewport Frame.
   2. Do **Zoom / All** before adjusting the scale.
   3. Select 1:1 in the **VIEWPORT** toolbar.

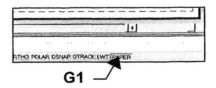

G1

G3

H. Lock the Viewport

This option will "lock" the adjusted scale within the selected viewport. If the viewport is "locked", the adjusted scale cannot be changed unless you "unlock" the viewport. When Zoom is used inside a locked viewport, the display of the entire layout is zoomed.

HOW TO LOCK A VIEWPORT

Method 1. a. Make sure you are in Paper Space.
            b. Click once on the Viewport frame
            c. Right Click (the short cut menu should appear)
            d. Select "Display Locked" - Select Yes or No.

Method 2. Double click on the Viewport frame. The "Properties Palette" will appear. Select Misc / Displayed Locked.

Method 3. Type **MV** at the command line and select "Lock".

J.   Save as: **My Decimal Setup**

K. Continue on to 3-D.

EXERCISE 3D
CREATING A PAGE SETUP

The following instructions will guide you through creating a **PAGE SETUP** named **"11X17 (1 to 1) All Black"**. You need this Page Setup to plot the 11x17 (1 to 1) layout. The Page Setup remembers which plotter to use, paper size, scale, and lineweights. This Page Setup will be embedded in **My Decimal Setup** and you will be able to use it over and over again. But you only have to create this page setup once.

A. Open **My Decimal Setup** if it isn't already open.

B. Select the **11x17 (1 to 1)** layout tab.

   You should be looking at your Border and Title Block now.

C. Select **File / Plot** or place the cursor on the **11x17 (1 to 1)** tab and press the right mouse button. Select **Plot** from the short cut menu.

   The Plot dialog box should appear

D. Select the **Plot Device** tab and select the options 1 thru 3 below:

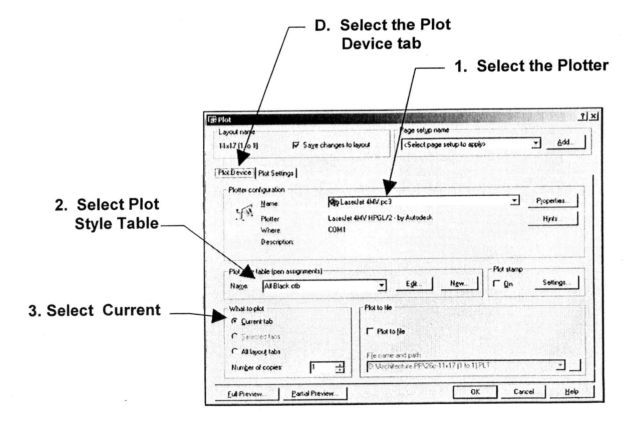

D. Select the Plot Device tab

1. Select the Plotter

2. Select Plot Style Table

3. Select Current

E. Select the **Plot Settings** tab then select options 1 thru 7 below:

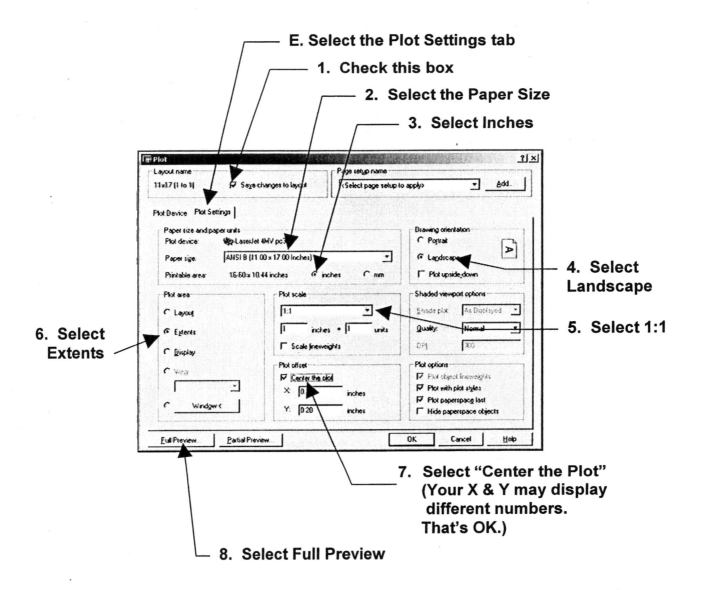

E. Select the Plot Settings tab

1. Check this box

2. Select the Paper Size

3. Select Inches

4. Select Landscape

5. Select 1:1

6. Select Extents

7. Select "Center the Plot"
(Your X & Y may display
 different numbers.
 That's OK.)

8. Select Full Preview

F. Preview the Layout

1. If the drawing is centered on the sheet, press the Esc key and continue
    on to **G**.

2. If the drawing does not look correct, press the Esc key and check all your
    settings, then preview again.

G. Select the **ADD** button.

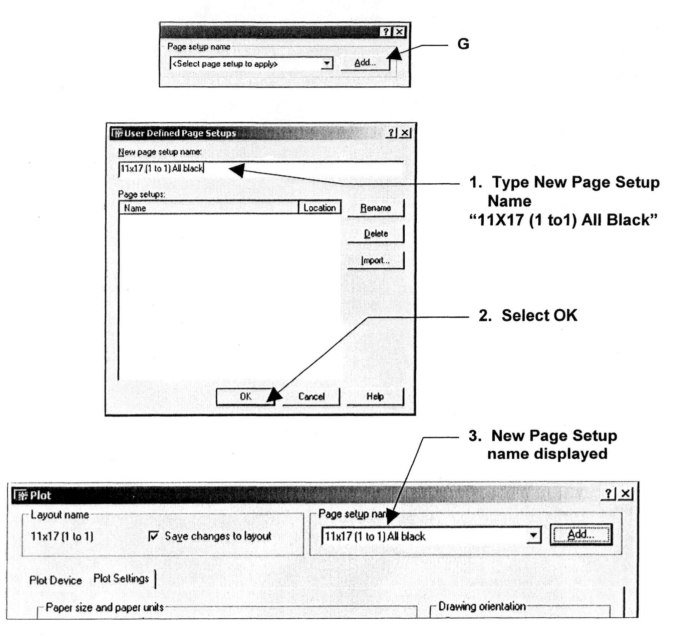

1. **Type New Page Setup Name**
    "11X17 (1 to1) All Black"

2. **Select OK**

3. **New Page Setup name displayed**

H. If your computer **is** connected to the Plotter / Printer, select the **OK** button to plot.

I. If your computer **is not** connected to the Plotter / Printer, select the **CANCEL** button. (Your settings **will not** be lost)

J. **Save** this file one more time:
    1. Select **File / Save** as
    2. Save as: **My Decimal Setup**

*You have now completed the **Page Setup** for the **My Decimal Setup**.   This Page Setup will be used to plot layout 11 X 17 (1 to 1).*

EXERCISE 3E
CREATE A LAYOUT & BORDER FOR PLOTTING ON A
24 X 18 SHEET

The following instructions will guide you through creating a new Layout and a new border to be used for plotting drawings on a 24 x 18 sheet.   All of this information will be saved and you will not have to do this again.

A. Open **My Decimal Setup**

B

B. Select the **Layout2** tab.

> *Note: If the Page Setup dialog box shown below does not appear automatically, right click on the Layout2 tab, then select "Page Setup". If you do not have a Layout2 tab, right click on the 11X17(1 to 1) tab and select "New Layout".*

1. Type the new name:
   24 X 18 (1 to 1)

2. Select the "*Plot Device*" tab.

3. Select the Plotter. (If this printer is not listed, see Appendix-A)

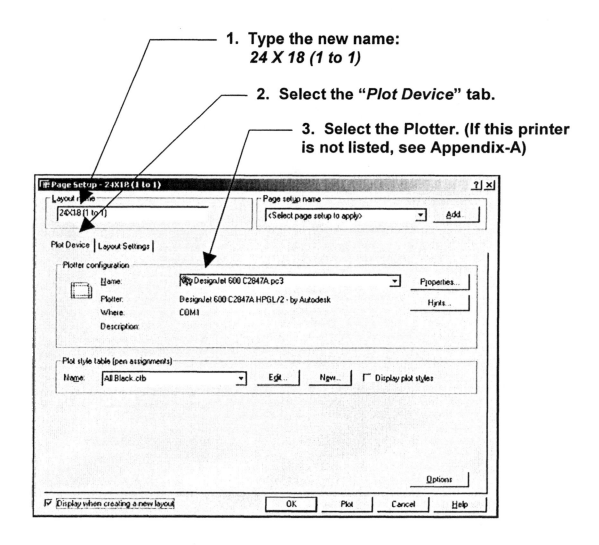

4. Select the "*Layout Settings*" tab.

5. Select the "*Paper Size*".

6. Select scale "1:1"

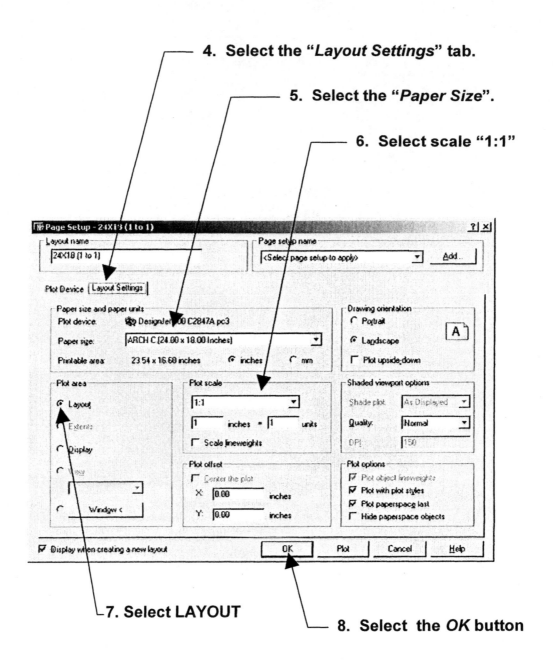

Page Setup - 24X18 (1 to 1)

Layout name
`24X18 (1 to 1)`

Page setup name
`<Select page setup to apply>`   Add...

Plot Device | **Layout Settings**

Paper size and paper units
Plot device: DesignJet 00 C2847A pc3

Paper size: `ARCH C (24.00 x 18.00 Inches)`

Printable area: 23.54 x 16.60 inches   ⊙ inches   ○ mm

Drawing orientation
○ Portrait
⊙ Landscape
□ Plot upside-down
[A]

Plot area
⊙ Layout
○ Extents
○ Display
○ View
○ Window <

Plot scale
`1:1`
`1` inches = `1` units
□ Scale lineweights

Plot offset
□ Center the plot
X: `0.00` inches
Y: `0.00` inches

Shaded viewport options
Shade plot   As Displayed
Quality: Normal
DPI: 150

Plot options
☑ Plot object lineweights
☑ Plot with plot styles
☑ Plot paperspace last
□ Hide paperspace objects

☑ Display when creating a new layout   OK   Plot   Cancel   Help

7. Select LAYOUT

8. Select the *OK* button

You should now have a sheet of paper displayed on the screen and the Layout2 tab should now be displayed as "24 X 18 (1 to 1)".

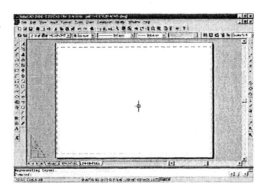

C. Now we need to draw a new larger RECTANGLE, shown below, on the sheet of paper shown on the screen. (Use Layer = Border)

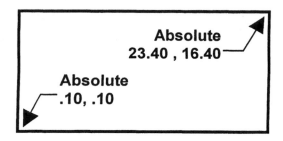

Absolute
23.40 , 16.40

Absolute
.10, .10

The next steps are to make a block from of the previous Title Block then you will Insert it into the 24 X 18 (1 to 1) layout. This is how AutoCAD saves you time.

D. Select the **"11 X 17 (1 to 1)"** Layout tab.

    1. You need to be in "Paperspace".
    2. Select **"DRAW / BLOCK / MAKE"**

        a. Make a BLOCK of the Title Block ONLY. Do not include the Border Rectangle. Notice where the "Basepoint" is selected below.

(Refer to "Exercise Workbook for Beginning AutoCAD 2004" Lesson 28 for step by step instructions for "Creating a Block" if you do not remember how)

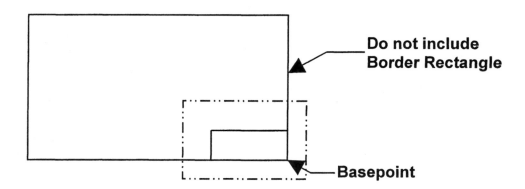

Do not include
Border Rectangle

Basepoint

E. Select the **24 X 18 (1 to 1)** Layout tab.
    1. **Very important:** Change to Layer = Border
    2. Select **INSERT / BLOCK**
    3. Select the "title block" block then select OK.
    4. Insertion Point should be the lower right corner of the Border Rectangle on the screen. (Refer to "Exercise Workbook for Beginning AutoCAD 2004" page 28-4 for step by step instructions)

Do not scale the inserted Block. It may look smaller, because the border rectangle is larger than the previous border, but the title block is the same size it was in 11 X 17 (1 to 1) layout.

F. Create a Viewport   (Refer to Exercise 3C for step by step instructions)

   1. Select Layer = Viewport
   2. Select "Single Viewport" icon.
   3. Draw a Viewport **approximately** 1/4" inside (smaller) the border lines.
       (Approximately as shown)

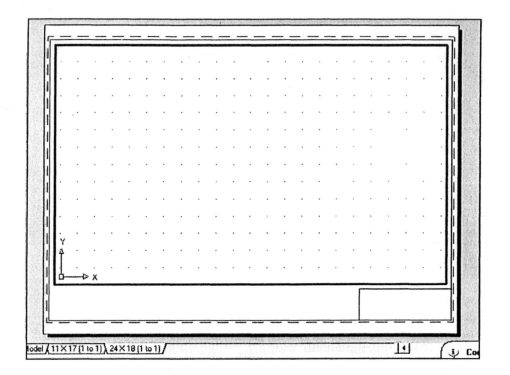

G. IMPORTANT: Change to **Model Space** before going on to **H**.
    (Double click inside the viewport or use the status line.)

H. Change the **DRAWING LIMITS**
       a. Lower left corner = 0 , 0
       b. Upper right corner = 24 , 18

I. Select **VIEW / ZOOM / ALL**

J. Select 1 : 1 in the **VIEWPORT toolbar.**

K. Lock the Viewport
    (Refer to page 3-11)

Now the Viewport scale will not change when you zoom in or out.

L. Save as: **My Decimal Setup (again)**

Now you have 2 master layout borders. One to be used when plotting on a sheet size 17 x 11 and one to be used when plotting on a sheet size 24 x 18 at a scale of 1 : 1.

EXERCISE 3F
CREATE A PAGE SETUP FOR 24 X 18 SHEET

The following instructions will guide you through creating a **PAGE SETUP** named:
"24 X 18 (1 to 1) All Black". You need this Page Setup to plot the **24 X18 (1 to 1)** layout. The Page Setup remembers which plotter to use, paper size, scale, and lineweights. This Page Setup will be embedded in **My Decimal Setup** and you will be able to use it over and over again. But you only have to create this page setup once.

A. Open **My Decimal Setup**

B. Select the **24 X 18 (1 to 1)** layout tab.

   You should be looking at your Border and title block

C. Select **File / Plot** or Place the cursor on the **24 X 18 (1 to 1)** tab, press the right mouse button and select "**Plot**" on the Short cut menu.

The Plot dialog box should appear

D. Select the **Plot Device** tab and then select options 1 thru 3 below:

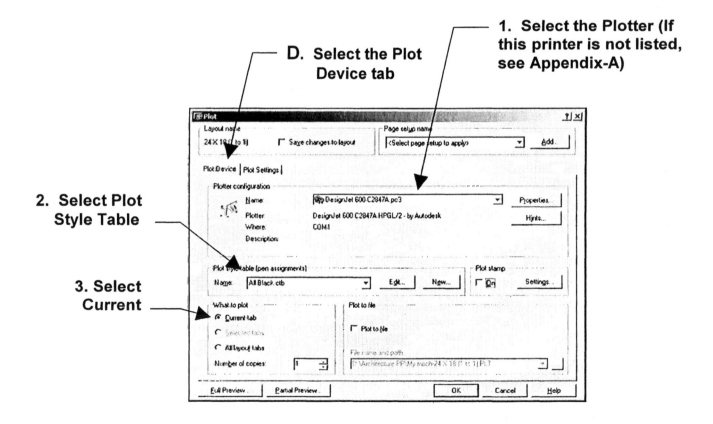

1. Select the Plotter (If this printer is not listed, see Appendix-A)

D. Select the Plot Device tab

2. Select Plot Style Table

3. Select Current

E. Select the **Plot Settings** tab then select options 1 thru 7 below:

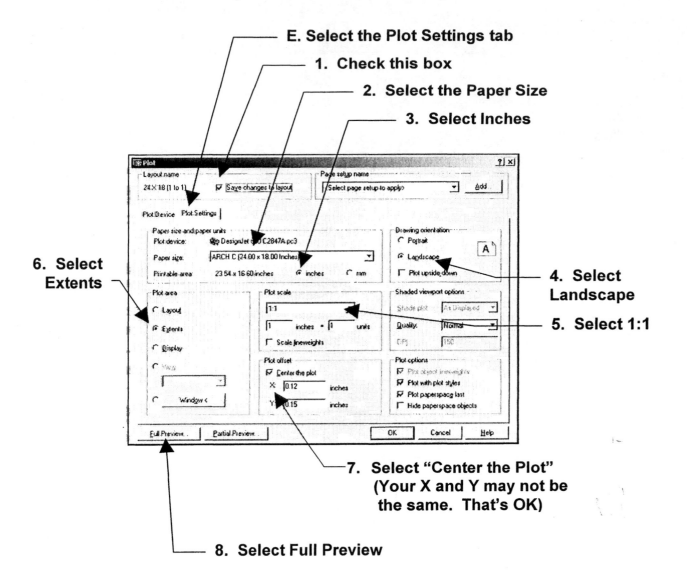

E. **Select the Plot Settings tab**

1. **Check this box**

2. **Select the Paper Size**

3. **Select Inches**

6. **Select Extents**

4. **Select Landscape**

5. **Select 1:1**

7. **Select "Center the Plot"**
   (Your X and Y may not be the same. That's OK)

8. **Select Full Preview**

F. Preview the Layout

   1. If the drawing is centered on the sheet, press the Esc key and go to **G**.

   2. If the drawing does not look correct, press the Esc key and check all your settings, then preview again.

G. Select the **ADD** button.

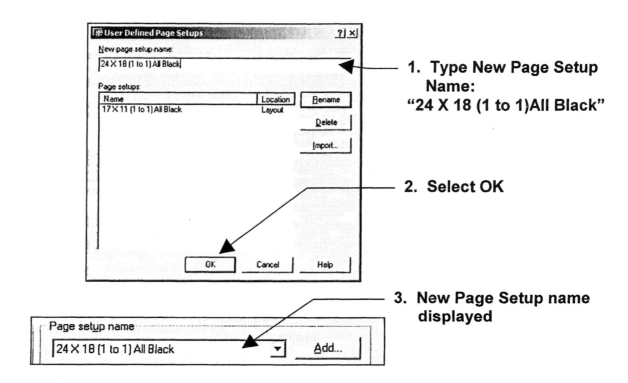

1. **Type New Page Setup Name:**
   "24 X 18 (1 to 1)All Black"

2. **Select OK**

3. **New Page Setup name displayed**

H. If your computer **is** connected to the Plotter / Printer, select the **OK** button to plot.

I. If your computer **is not** connected to the Plotter / Printer, select the **CANCEL** button. (Your settings **will not** be lost)

J. **Save** this file <u>one more time</u>:
   1. Select **File / Save** as
   2. Save as: **My Decimal Setup (again)**

*You have now created another **Page Setup** for the **My Decimal Setup**. This Page Setup will be used to plot layout 24 X 18 (1 to 1).*

EXERCISE 3G
CREATE A LAYOUT & BORDER FOR PLOTTING ON A
24 X 36 SHEET

The following instructions will guide you through creating a new Layout and a new border to be used for plotting drawings on a 24 x 36 sheet.   All of this information will be saved with My Decimal Setup and you will not have to do this again.

A. Open **My Decimal Setup**

B. **Right Click** on the **24 X 18 (1 to 1)** layout tab.
1. Select **"NEW LAYOUT"** from the short cut menu.
(A new layout1 tab will appear)

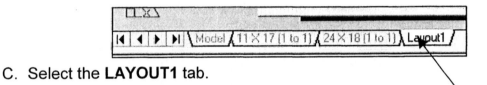

C. Select the **LAYOUT1** tab.

Note: If the Page Setup dialog box shown below does not appear automatically, right click on the Layout1 tab, then select Page Setup.

1. **Type the new name:**
   24 X 36 (1 to 1)

2. **Select the "*Plot Device*" tab.**

3. **Select the Plotter. (If this printer is not listed, see Appendix-A)**

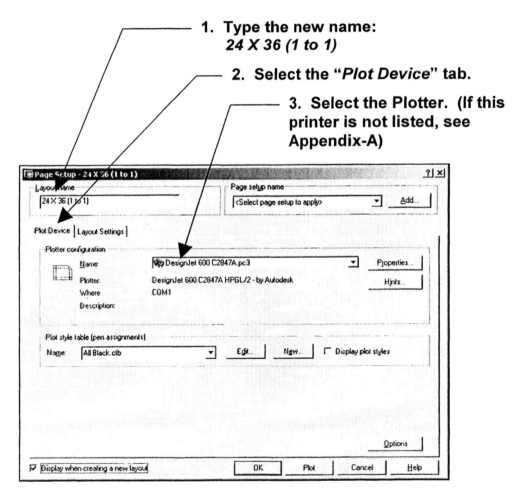

4. Select the "*Layout Settings*" tab.

5. Select the "*Paper Size*".

6. Select scale "1:1"

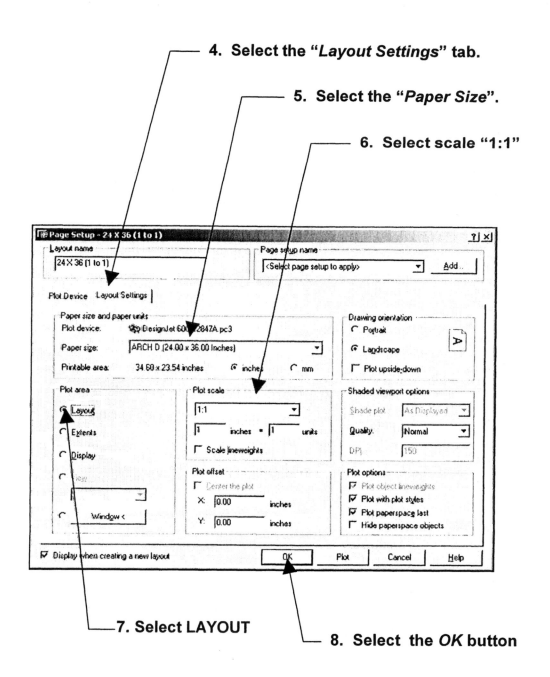

7. Select LAYOUT

8. Select the *OK* button

You should now have a sheet of paper displayed on the screen and the Layout1 tab should now be displayed as "24 X 36 (1 to 1)".

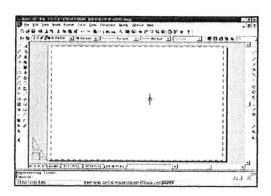

D. Draw the **RECTANGLE** below. (Use Layer = Border)

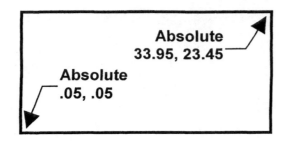

E. Insert the **TITLE BLOCK** block just like you did in the 24 X 18 (1 to 1) Layout.

F. Create a **Viewport**
   1. You decide what the coordinates should be. Just using the mouse is OK but keep the viewport close and inside the border rectangle.

G. Change to **Model Space**. (double click inside the Viewport)

H. Change the **DRAWING LIMITS**
   a. Lower left corner = 0 , 0
   b. Upper right corner = 36 , 24

I. IMPORTANT: Select **VIEW / ZOOM / ALL**

J. Select 1: 1 in the Viewport toolbar.

K. Lock the Viewport
   (Refer to 3-11)

Now the Viewport scale will not change when you zoom in or out.

L. Save as: **My Decimal Setup**

Now you have 3 master layout borders. One to be used when plotting on a sheet size 17 x 11, one to be used when plotting on a sheet size 24 x 18 and one to be used when plotting on a sheet size 24 x 36 and all at a scale of 1 : 1.

Now you are almost done. Just one more step.

EXERCISE 3H
CREATE A PAGE SETUP FOR 24 X 36 SHEET

The following instructions will guide you through creating a **PAGE SETUP** named:
"24 X 36(1 to 1) All Black". You need a Page Setup to plot the **24 X 36 (1 to 1)** layout.
The Page Setup remembers which plotter to use, paper size, scale, and lineweights.
This Page Setup will be embedded in **My Decimal Setup** and you will be able to use it
over and over again. But you only have to create this page setup once.

A. Open **My Decimal Setup**

B. Select the **24 X 36 (1 to 1)** layout tab. (You should be looking at your Border and
    title block)

C. Select **File / Plot** or Place the cursor on the **24 X 36 (1 to 1)** tab, press the right
    mouse button then select "**Plot**" in the Short cut menu.

   The Plot dialog box should appear

D. Select the **Plot Device** tab and select the options 1 thru 3 below:

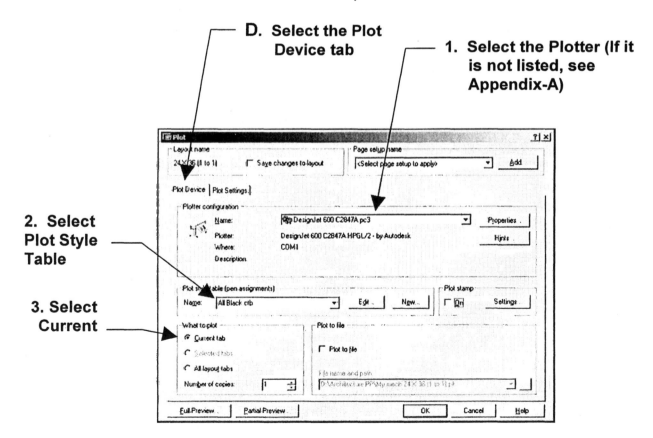

D. Select the Plot Device tab

1. Select the Plotter (If it is not listed, see Appendix-A)

2. Select Plot Style Table

3. Select Current

E. Select the **Plot Settings** tab then select options 1 thru 8 below:

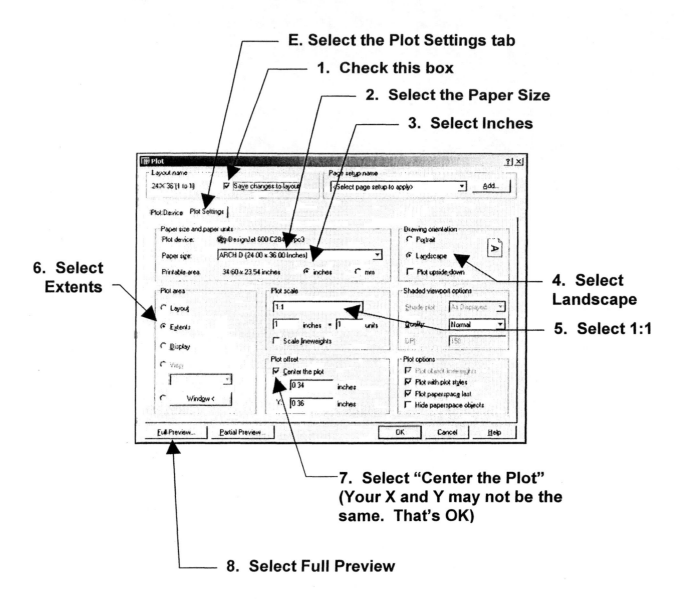

E. Select the Plot Settings tab

1. Check this box

2. Select the Paper Size

3. Select Inches

6. Select Extents

4. Select Landscape

5. Select 1:1

7. Select "Center the Plot" (Your X and Y may not be the same. That's OK)

8. Select Full Preview

F. Preview the Layout

1. If the drawing is centered on the sheet, press the Esc key and go to **G**.

2. If the drawing does not look correct, press the Esc key and check all your settings, then preview again.

G. Select the **ADD** button.

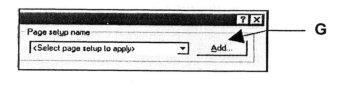

G

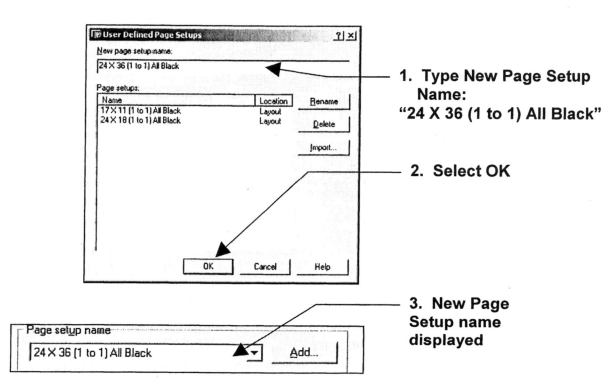

1. **Type New Page Setup Name:**
   "24 X 36 (1 to 1) All Black"

2. **Select OK**

3. **New Page Setup name displayed**

H. If your computer **is** connected to the Plotter / Printer, select the **OK** button to plot.

I. If your computer **is not** connected to the Plotter / Printer, select the **CANCEL** button.
   (Your settings will not be lost)

J. **Save** this file one more time:
   1. Select **File / Save** as
   2. Save as: **My Decimal Setup**

You have now created yet another **Page Setup** for the *My Decimal Setup*. This page setup will be used with the 24 X 36 (1 to 1) layout.

Summary

Now let's think about what you have achieved in this lesson.

3A.   Select Preferences to draw with.

1. Started a NEW drawing file.
2. Selected your preferences such as: Units, Drawing Limits, Snap and Grid.
3. Created New Layers, Text Styles and a Dimension Style.
4. Saved the file as: My Decimal Setup

3B.   Select Paper size to plot on.

1. Selected the "Size" of sheet you want to plot on.
2. Drew a master border with title block.
3. Saved the file as: My Decimal Setup

3C.   Cut a hole in the sheet to see model space and adjust it's scale.

1. Create a Viewport
2. Adjusted the scale of the Viewport
3. Locked the Viewport
4. Saved the file as: My Decimal Setup

3D.   Set up Plotting specs and save them.

1. Select Plotter
2. Select the "color dependent plot style"
3. Select what size paper
4. Select what area to plot.
5. Select the scale for paperspace.
6. Select the position, such as "centered".
7. Assign a name for these specs
8. Add the specs to your file.

3E.   Repeat 3B and 3C for a larger size of paper.

3F.   Repeat 3D with different specs

3G.   Repeat 3B and 3C for an even larger size of paper.

3H.   Repeat 3D with different specs.

The result is a Master setup drawing file with 3 layout tabs predefined for 3 sizes of paper and 3 page setups (plotting specifications) for each layout.

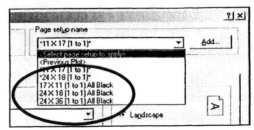

LEARNING OBJECTIVES

After completing this lesson, you will be able to:

1. Create a "Feet and inches" Set up drawing.
2. Create another Border for 24 x 18
3. Create Page Set ups for 1/4" = 1' and 1/8" = 1'

LESSON 4

EXERCISE 4A
CREATE A MASTER FEET-INCHES SETUP DRAWING

The following instructions will guide you through creating a "Master" feet-inches setup drawing.

NEW SETTINGS

A. Begin your drawing without a template as follows:

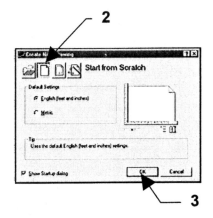

1. Select **"FILE / NEW"**
2. Select **"START FROM SCRATCH"** Box.
3. Select **"OK"**.
4. Your screen should be blank, no grids and the current layer is 0

B. Set drawing specifications as follows:

1. Set **"UNITS"** of measurement
   Use "**FORMAT / UNITS** and change the settings as shown below, then select **OK.**

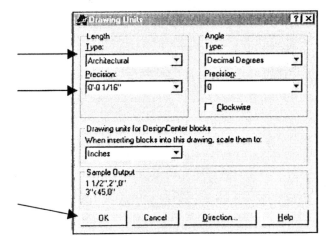

2. Set **"DRAWING LIMITS"** (Size of drawing area)
   USE "**FORMAT / DRAWING LIMITS**
   a. Lower left corner = 0'-0", 0'-0"
   b. Upper right corner = 68', 44' (Notice this is feet not inches)
   c. Use **"VIEW / ZOOM / ALL"** to generate the new limits
   d. Set your **grids** to **ON** to display the paper size.

3. Set **"SNAP AND GRID"**
    Use **"TOOLS / DRAFTING SETTINGS"**

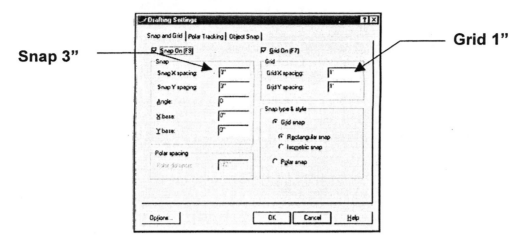

Snap 3"

Grid 1"

4. Set **"PICK BOX"** size (your preference)
    Use "Tools / Options / Selection" Tab

NEW LAYERS

C. Create new layers

    1. **Load** linetype
        DASHED2
        PHANTOM

    2. Assign names, colors, linetypes and plotability. (Use "Format / Layer")

NAME	COLOR	LINETYPE	LWT	PLOT
BORDER	RED	CONTINUOUS	.039	YES
CABINETS	CYAN	CONTINUOUS	.016	YES
CONSTRUCTION	WHITE	CONTINUOUS	Default	NO
DIMENSION	BLUE	CONTINUOUS	Default	YES
DOORS	GREEN	CONTINUOUS	.016	YES
ELECTRICAL	CYAN	CONTINUOUS	.010	YES
FURNITURE	MAGENTA	CONTINUOUS	.016	YES
HATCH	GREEN	CONTINUOUS	Default	YES
HIDDEN	MAGENTA	DASHED2	Default	YES
MISC	CYAN	CONTINUOUS	.016	YES
OBJECT	RED	CONTINUOUS	.024	YES
PLUMBING	9	CONTINUOUS	.010	YES
PROPERTYLINE	WHITE	PHANTOM	.024	YES
SYMBOLS	GREEN	CONTINUOUS	.024	YES
TEXT HEAVY	WHITE	CONTINUOUS	Default	YES
TEXT LIGHT	BLUE	CONTINUOUS	Default	YES
VIEWPORT	GREEN	CONTINUOUS	Default	NO
WALLS	RED	CONTINUOUS	.024	YES
WINDOWS	GREEN	CONTINUOUS	.016	YES
WIRING	CYAN	DASHED	.010	YES
XREF	WHITE	CONTINUOUS	Default	YES

NEW TEXT STYLE

D. Create a 2 text styles named *ARCH TEXT* and *CLASS TEXT*

    1. Select "**FORMAT / TEXT STYLE**"
    2. Make the changes shown in the dialog boxes below.

ARCH TEXT	CLASS TEXT

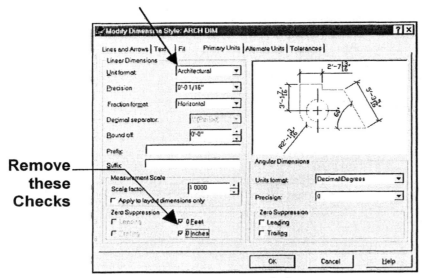

    3. When complete, select **APPLY** then **CLOSE**.

NEW DIMENSION STYLE

E. Create a new Dimension Style named *Arch Dim*
(Refer to Exercise Workbook for Beg. AutoCAD 2002, Lesson 16-6)

1. Select **FORMAT / DIMENSION STYLE**
2. Select the **NEW** button.
3. Make the changes to the following dialog
boxes.

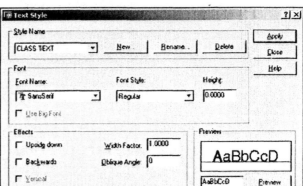

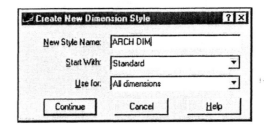

Remove these Checks

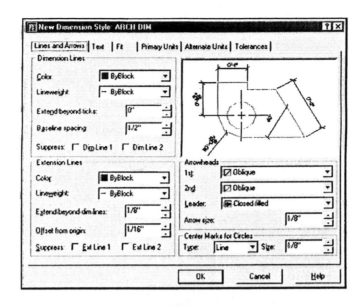

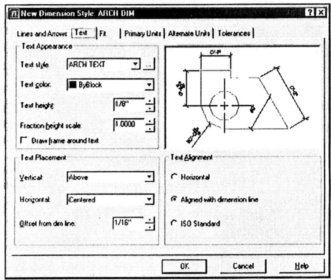

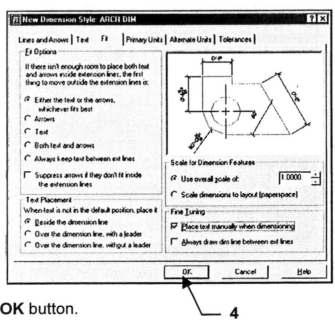

4. **NOW** select the **OK** button.

4

5. *Your new style "**ARCH DIM**" should be listed.* Select the **"Set Current"** button to make your new style "**ARCH DIM**" the style that you will use.

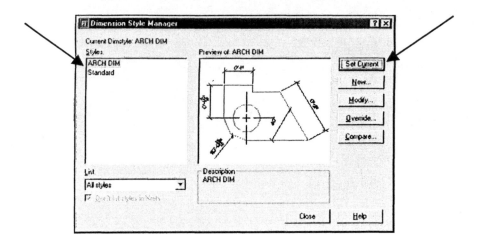

6. Select the **Close** button

THIS NEXT STEP IS VERY IMPORTANT.

F. SAVE ALL THE SETTINGS YOU JUST CREATED
      1. Select   File / Save as
      2. Save as: **My Feet-Inches Setup**

NOTE: This was just step 1 to creating your setup drawing. Continue on through the following exercises, in this lesson, using this same file (My Feet-Inches Setup). Each step adds more information to this drawing.

EXERCISE 4B
CREATE AN ARCHITECTURAL BORDER FOR PLOTTING

The following instructions will guide you through creating a Border drawing that will be used in combination with "My Feet-Inches Setup" when plotting. You will create a Layout and draw a border with a title block. All of this information will be saved and you will not have to do this again.

A. Open **My Feet-Inches Setup**

B. Select the **LAYOUT1** tab.

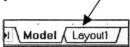

Note: If the Page Setup dialog box shown below does not appear automatically, right click on the Layout tab, and then select Page Setup from the short cut menu.

1. **Type the new name:**
   24 X 18 Full

2. **Select the "*Plot Device*" tab.**

3. **Select the Plotter.**

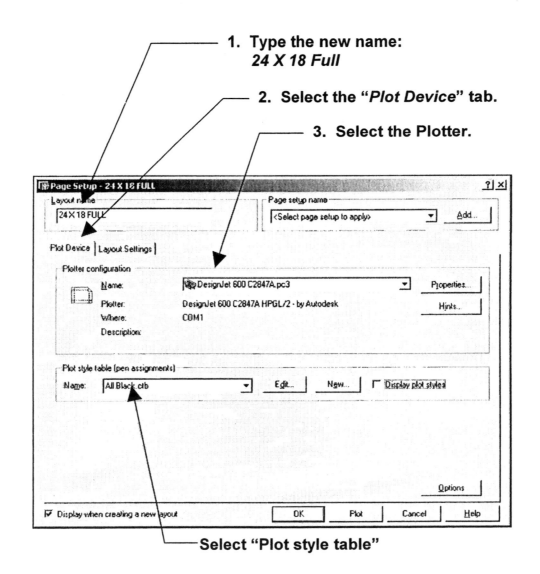

Select "Plot style table"

5. Select the "*Layout Settings*" tab.

6. Select the "*Paper Size*".

7. Select scale "1:1"

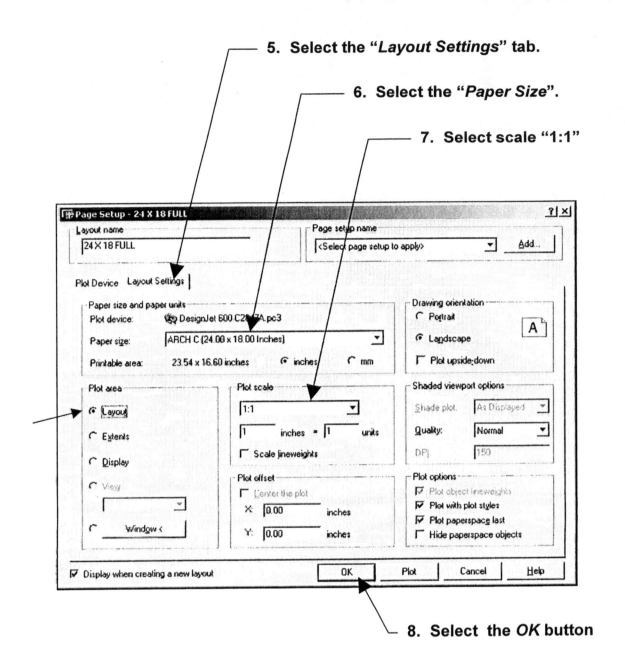

8. Select the *OK* button

You should now have a sheet of paper displayed on the screen and the Layout tab should now be displayed as **"24 X 18 Full"**.

This sheet is in front of "Model". In Exercise 4D, you will cut a hole in this sheet so you can see through to Model.

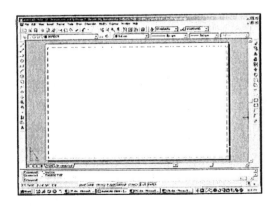

C. Draw the Border with title block, shown below, on the sheet of paper shown on the screen.

D. When you have completed the Border, shown below:
1. Select File / Save as
2. Save as: **My Feet-Inches setup** (Again)

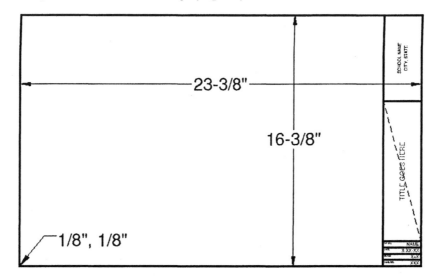

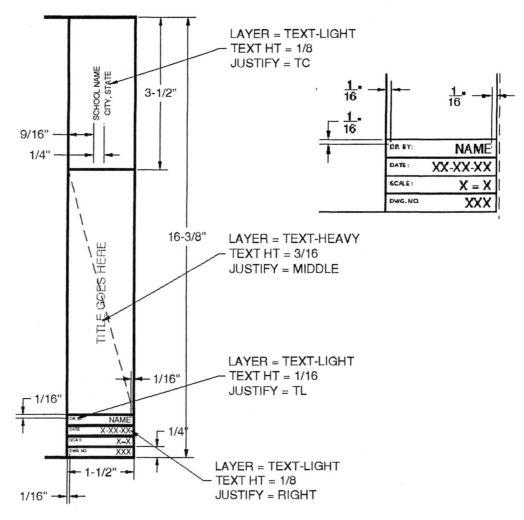

Don't forget to save, Refer to "D" above.

EXERCISE 4C
CREATE A VIEWPORT

The following instructions will guide you through creating a VIEWPORT in the Border Layout sheet. Creating a viewport has the same effect as cutting a hole in the sheet of paper. You will be able to see through the hole to Model.

A. Open **My Feet-Inches Setup**

B. Select the **24 X 18 Full** tab.

C. Select layer **"Viewport"**

D. Select the **Single Viewport icon** from the Viewports Toolbar.

— Single Viewport icon

E. Draw a single viewport approximately 1/8" inside (smaller) the border lines, as shown.

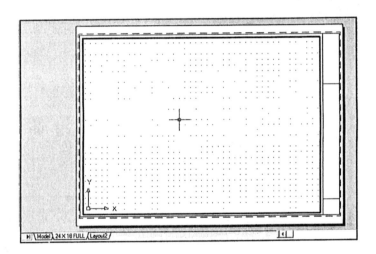

F. After successfully creating the Viewport, you should now be able to see through to Model. (Your grids will appear if they are ON. Turn ON in model space, OFF in Paper space)

G. Adjust the Model space scale.
   1. Change the **PAPER** button to **Model** or double click inside the Border area.
   2. Select 1:1 in the **VIEWPORT** toolbar.

> **Note: When you are using <u>Architectural units</u> and you select <u>1:1</u>, AutoCAD displays <u>1" = 1'</u>. This is an error in the AutoCAD software. It should display 1" = 1". But don't worry, it will function as <u>1 : 1</u>.**

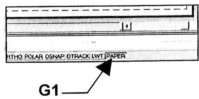
G1

H. **Lock** the Viewport. (Refer to 3-11 if necessary)

I. File / Save as: **My Feet-Inches Setup (again)**

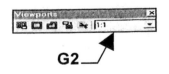
G2

EXERCISE 4D
CREATE A PAGE SETUP FOR 24 X 18 SHEET

The following instructions will guide you through creating a **PAGE SETUP** named: **"24 X 18-All Black".** You need a Page Setup to plot the **24 X 18 Full** layout. The Page Setup remembers which plotter to use, paper size, scale, and lineweights. This Page Setup will stay with **My Feet-Inches Setup** and you will be able to use it over and over again. But you only have to create this page setup once.

A. Open **My Feet-Inches Setup**

B. Select the **24 X 18 Full** layout tab.

   You should be looking at your Architectural Border now.

C. Select **File / Plot** or Place the cursor on the **24 X 18 Full** tab, press the right mouse button, then select **Plot** from the Short cut menu.

   The Plot dialog box should appear

D. Select the **Plot Device** tab and select the options 1 thru 3 below:

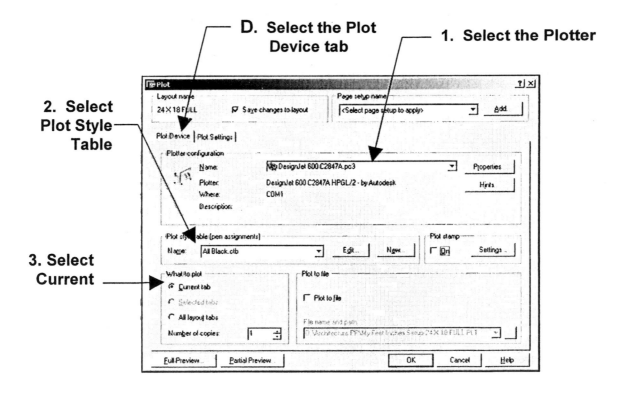

Do Not select OK yet. Go on to "E" on the next page.

E. Select the **Plot Settings** tab then select options 1 thru 7 below:

E. Select the Plot Settings tab

1. Check this box

2. Select the Paper Size

3. Select Inches

6. Select Extents

4. Select Landscape

5. Select 1:1

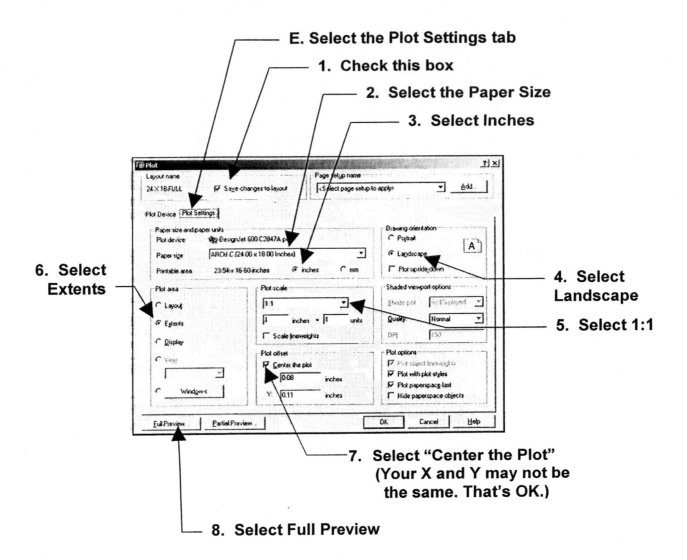

7. Select "Center the Plot"
(Your X and Y may not be
the same. That's OK.)

8. Select Full Preview

F. Preview the Layout

1. If the drawing is centered on the sheet, press the Esc key and go to **G**.

2. If the drawing does not look correct, press the Esc key and check all your settings, then preview again.

G. Select the **ADD** button.

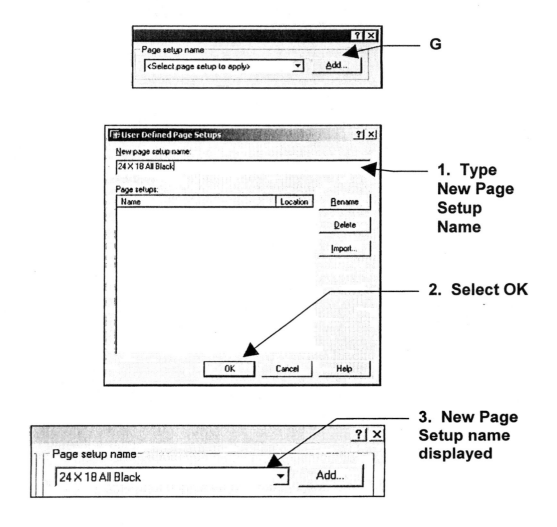

             G

1. Type New Page Setup Name

2. Select OK

3. New Page Setup name displayed

H. If your computer **is** connected to the Plotter / Printer, select the **OK** button to plot.

I. If your computer **is not** connected to the Plotter / Printer, select the **CANCEL** button. (Your settings **will not** be lost)

J. **Save** this file one more time:
    1. Select **File / Save** as
    2. Save as: **My Feet-Inches Setup**

You have now created a **Page Setup** for the ***My Feet-Inches Setup***. This page setup will be used to plot all layouts, in this setup drawing, that use 24 X 18 sheet size.

EXERCISE 4E

Create an architectural layout & border for plotting on a 24 x 18 sheet and the Model Space drawing scale will be adjusted to ¼" = 1'

The following instructions will guide you through creating a Border drawing that will be used when your drawing gets so large that you need to plot them at a scale of ¼" = 1'. You will create a new Layout and **INSERT** the previously drawn border with title block. You will then cut a viewport and adjust the scale of the Model space to ¼" = 1'. This will not be difficult. Just follow the directions below carefully. All of this information will be saved and you will not have to do this again.

A. Open **My Feet-Inches Setup**

B. Select the 24X18 FULL tab.
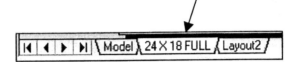

C. Make a **BLOCK** of the border rectangle and title block.
1. Use **DRAW / BLOCK / MAKE**
2. Name = **ARCH BORDER**
3. Base Point = 0, 0, 0

D. Select the **LAYOUT2** tab.
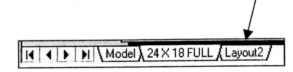

Note: If the Page Setup dialog box shown below does not appear automatically, right click on the Layout2 tab, then select Page Setup from the short cut menu.

1. **Type the new name: *Qtr Equals Foot***

2. **Select the "*Plot Device*" tab.**

3. **Select the Plotter.**

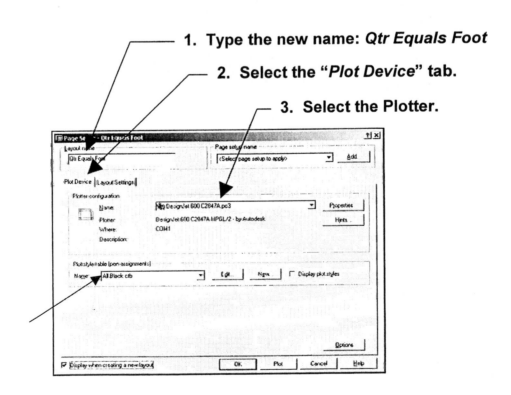

4. Select the "*Layout Settings*" tab.

5. Select the "*Paper Size*".

6. Select scale "1:1"

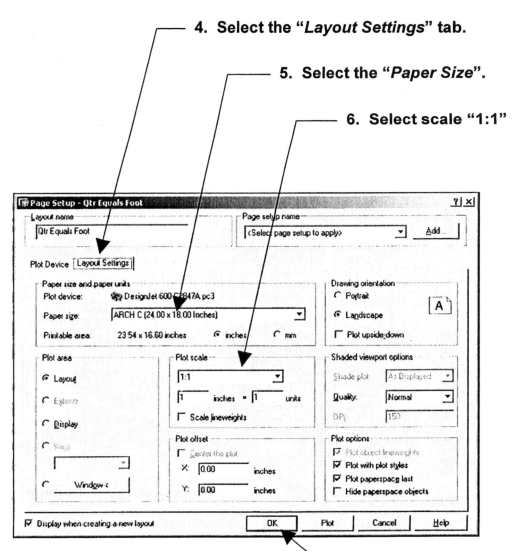

7. Select the *OK* button
(Your X and Y may not
be the same. That's OK.)

You should now have a sheet of paper displayed on the screen and the Layout2 tab should now be displayed as "Qtr Equals Foot".
This sheet is in front of "Model".

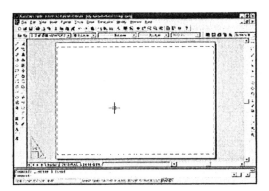

E. **Insert** the Block (**ARCH BORDER**).
   1. Change to Layer = Border
   2. Use **INSERT / BLOCK**
   3. Insertion Point = 0,0,0

Note: __Explode__ and __Erase__ the old viewport frame before going on to "F". It is no longer a viewport. It is just a rectangle.

Single
Viewport icon

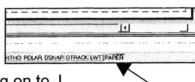

F. Select layer **"Viewport "**
G. Select the **Single Viewport icon** from the Viewports Toolbar.
H. Draw a single viewport approximately 1/8" inside (smaller) the border lines, as shown.

After successfully creating the Viewport, you should now be able to see through to Model. (Your grids may not appear yet)

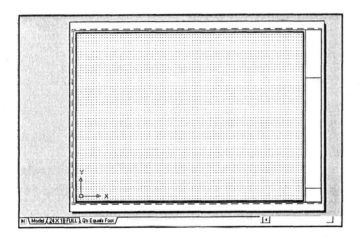

I. IMPORTANT: Change to **MODEL SPACE** before going on to J.
(Double click in drawing area or select the Paper button.)

J. Change the **DRAWING LIMITS**
   a. Lower Left corner = **0, 0**
   b. Upper Right corner = **96' , 72' (This is feet not inches)**

K. Select **VIEW / ZOOM / ALL**

L. Adjust the Model space scale to ¼" = 1'.

M. Change the **Grid** and **Snap** settings (Tools / Drafting Settings)
   1. Snap = 3"   Grids = 1'
   2. If your grids do not appear, make sure they are ON. (ON in model, Off in paper)

N. **Lock** the Viewport.

O. File / Save as again: **My Feet-Inches Setup**

P. Note: Use your previously created Page Setup **"24 X18 All Black"** when plotting this Layout.

EXERCISE 4F

Create an architectural layout & border for plotting on a 24 x 18 sheet and the Model Space drawing scale will be adjusted to 1/8" = 1'

The following instructions will guide you through creating a Border drawing that will be used when your drawings get so large that you need to plot them at a scale of 1/8" = 1'. You will create a new Layout and **INSERT** the previously drawn border with title block. You will then cut a viewport and adjust the scale of the Model space to 1/8" = 1'. This will not be difficult. Just follow the directions below carefully. All of this information will be saved and you will not have to do this again.

A. Open **My Feet-Inches Setup**

B. Create another Layout tab.
1. <u>Right Click</u> on the **"Qtr Equals Foot"** tab and select **New layout** from the short cut menu.

C. Select the new **LAYOUT1** tab.

|◄|◄|►|►|\ Model ⟨ 24 X 18 FULL ⟨ Qtr Equals Foot ⟩ Layout1 /

Note: If the Page Setup dialog box shown below does not appear automatically, right click on the Layout1 tab, then select Page Setup from the short cut menu.

1. Type the new name: "*Eighth Equals Foot*"

2. Select the "*Plot Device*" tab.

3. Select the Plotter.

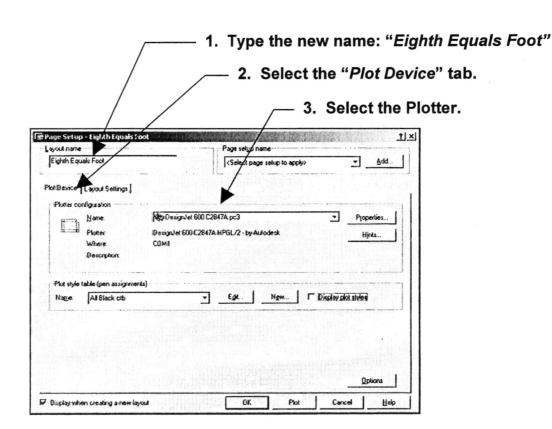

4. Select the "*Layout Settings*" tab.

5. Select the "*Paper Size*".

6. Select scale "1:1"

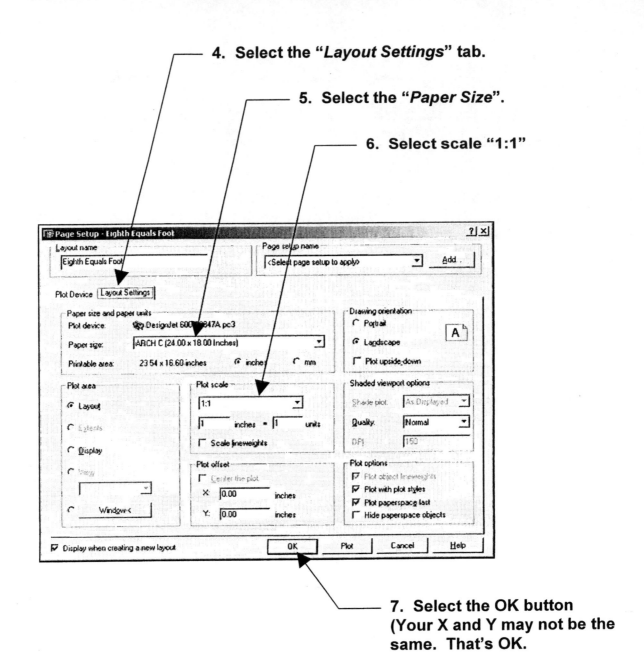

7. Select the OK button (Your X and Y may not be the same. That's OK.

You should now have a sheet of paper displayed on the screen and the Layout1 tab should now be displayed as **"Eighth Equals Foot"**.
This sheet is in front of "Model".

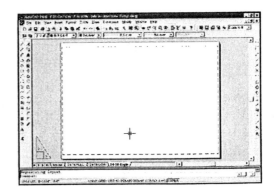

D. **Insert** the Block (**ARCH BORDER**).
   1. ***Important***, First change to Layer Border
   2. Use **INSERT / BLOCK**
   3. Insertion Point = 0,0,0

Note: <u>EXPLODE</u> and <u>Erase</u> the old viewport frame before going on to "F". It is no longer a viewport. It is just a rectangle.

Single Viewport icon

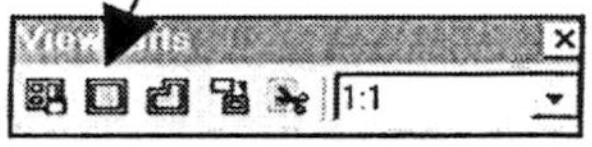

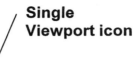

E. Select layer **"Viewport "**
F. Select the **Single Viewport icon** from the Viewports Toolbar.
G. Draw a single viewport approximately 1/8" inside (smaller) the border lines, as shown.

After successfully creating the Viewport, you should now be able to see through to Model. (Your grids may not appear yet)

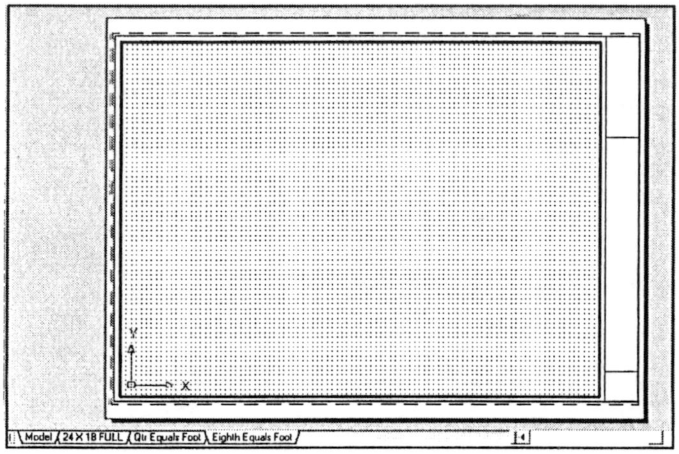

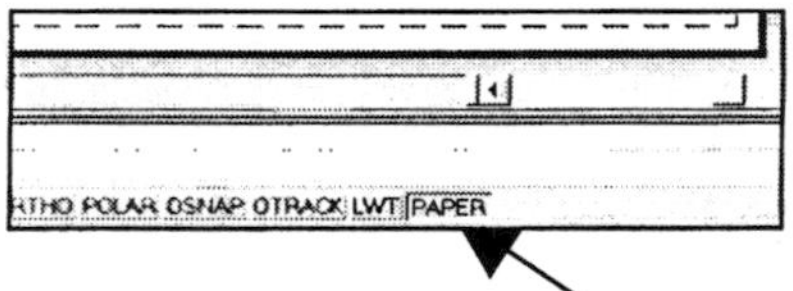

H

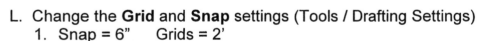

H. IMPORTANT: Change to **MODEL SPACE** before going on to J.

I. Change the **DRAWING LIMITS**
   a. Lower Left corner = **0, 0**
   b. Upper Right corner = **192' , 144' (Feet not inches)**

J. Select **VIEW / ZOOM / ALL**

K. Adjust the Model space scale to 1/8" = 1'

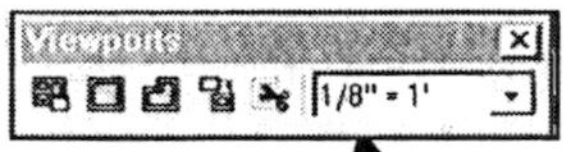

K

L. Change the **Grid** and **Snap** settings (Tools / Drafting Settings)
   1. Snap = 6"   Grids = 2'

M. **Lock** the Viewport.

N. File / Save as again: **My Feet-Inches Setup**

O. Note: Use your previously created Page Setup **"24 X18 All Black"** when plotting this Layout.

Summary

Let's review what you achieved in this lesson.

4A. **Select Preferences to draw with.**
1. Specified Units, Drawing Limits, Snap and Grid.
2. Created New Layers, Text Styles and a Dimension Style.

4B. **Select Paper Size 24 X 18 to plot on and drew a Border and Title Block.**

4C. **Created a Viewport, Adjusted the Scale of Model Space and Locked the Viewport.**

4D. **Set up Plotting specs (Page Setup) and saved them. (24 X 18 All Black)**
This Page Setup can be used to plot any Layout tab in this Master Setup drawing, as long as it is plotted on 24 X 18 sheet size.

4E/F. **Create a new Layout with the scale within the Viewport scaled to 1/4" =1'.**
It is good drawing management to setup Layouts already prepared to use when plotting with different scales.
4C created a layout for plotting model space at a scale of 1 : 1.
4E created a layout for plotting model space at a scale of !/4" = 1'
4F created a layout for plotting model space at a scale of 1/8" = 1'

Layouts are merely a process of preparing your drawing for printing. You may have up to 16 Layouts. In this lesson you have been shown how to create a master setup drawing to use when you draw in Architectural units. The Layouts already have the scale adjusted for plotting 1 : 1, 1/4" = 1' or 1/8" = 1'

*Notice that you only need one page setup (24 X 18 All Black) to plot all 3 Layouts. But the previous lesson 3 you needed 3 different page setups. Why? Because you need a page setup for each **sheet size**. In lesson 3, we are using 3 different sheet sizes. 17 X 11, 24 X 18 and 24 X 36. In this lesson, we are only using one sheet size, 18 X 24.*

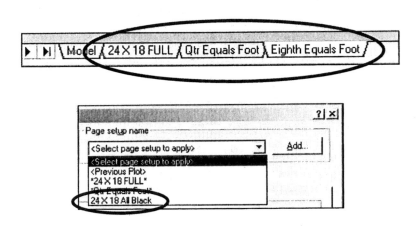

LEARNING OBJECTIVES

After completing this lesson, you will be able to:

1. Create a Multiline Style
2. Draw a multiline
3. Edit a Multiline
4. Create and use Layer States

LT users only
5. Double Line

LESSON 5

MULTILINE   (Not available in LT, refer to page 5-8)

The **MULTILINE** command is a method for drawing a set of parallel lines. The set can have up to 16 parallel lines. These multilines can basically be drawn in the same manner as a line, with first and second endpoints.   You can even use the CLOSE option. But the set is only **one** object, not multiple objects. The walls on a Floorplan would be an example of an Architectural application. The belts on a conveyor would be an example of a Mechanical application. Multiple circuit lines on a schematic would be an example of an Electrical application.

Before using the MULTILINE command you must "**Create a MULTILINE STYLE**".
(Or *you may merely want to use the AutoCAD default style named STANDARD.*)
Instructions for "Creating a Multiline Style" are on page 5-4.

First we will discuss how to draw with the MULTILINE command using the AutoCAD default Multiline Style "Standard".

1. Select the MULTILINE command using one of the following commands:

> **TYPE = ML**
> **PULLDOWN = DRAW / MULTILINE**
> **TOOLBAR = DRAW**   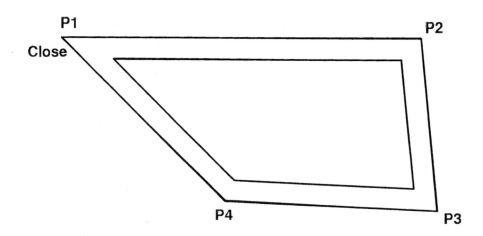 Multiline

> Command: _mline
> Current settings: Justification = Top, Scale = 1.00, Style = STANDARD

2. Specify start point or [Justification/Scale/STyle]: *select the start point (P1)*
3. Specify next point: *select the next point (P2)*
4. Specify next point or [Undo]: *select the next point (P3)*
5. Specify next point or [Close/Undo]: *select the next point (P4)*
6. Specify next point or [Close/Undo]: *Close <enter>*

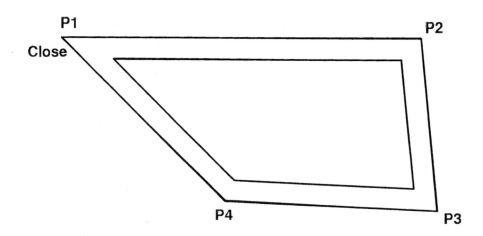

JUSTIFICATION

Because there are multiple lines, you must specify the justification of the multiline. Basically, this means: *what is your cursor hanging on to*? Is it the top, middle or bottom of the set, while you specify the endpoints. Your choices are **Top, Zero and Bottom.** These justifications are defined for drawing the multiline in a **Clockwise** direction.

1. Select the Multiline command
2. Select Justification option
3. Select Top, Zero or Bottom

Examples of Justification (drawn clockwise):

TOP   **ZERO**   **BOTTOM**

SCALE

The SCALE option is a **scale factor** of the width. It will increase or decrease the original specified width by multipling the original width by the specified scale.

1. Select the Multiline command
2. Select the Scale option
3. Enter the multiline scale

STYLE

The STYLE option allows you to select a previously created multiline style from the command line rather than having to select the FORMAT / MULTILINE pulldown menu.

1. Select the Multiline command
2. Select the Style option
3. Enter the Multiline Style name.

CREATING A MULTILINE STYLE

The following instructions will guide you through creating a new multiline style. The order in which you define your specifications is very important. Do not jump around, follow the order below.

Select the Multiline Style command using one of the following commands:

TYPE = MIstyle
PULLDOWN = FORMAT / MULTILINE STYLE
TOOLBAR = NONE

The following dialog box will appear.

4. Select the SAVE button if you want to save the style to use on other drawings.

5. Select OK when the settings are complete.

1. Type the new multiline style name here.

2. Select the ADD button to add the style to your drawing.

3. Set the ELEMENT PROPERTIES and MULTILINE PROPERTIES (Explanation shown below)

ELEMENT PROPERTIES

Each individual line within the MULTILINE set has it's own offset, linetype and color. These are called line ELEMENTS.

1. **OFFSET** is the distance from the center of the multiline. The default offset shown is .50 above the center and .50 below.
   a. Highlight the line you want to change
   b. Make the change in the offset box.

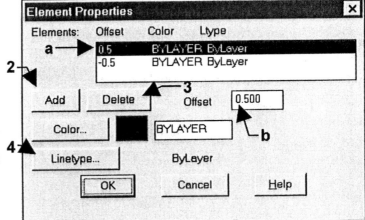

2. **ADD** The add button is used to add a line to the multiline set.
   a. Select the add button. A line will appear with an offset of 0.
   b. Change the offset distance.

3. **DELETE** The delete button is used to delete a line from the multiline set.
   a. Select the line you want to delete
   b. Select the delete button.

4. **COLOR / LINETYPE** When you select either of these buttons, the standard color of linetype dialog box appears. Select the desired color or linetype then OK.

MULTILINE PROPERTIES

This dialog box controls the end capping, fill and joint display properties of the multiline style. The capping can be applied at the start, end or both.

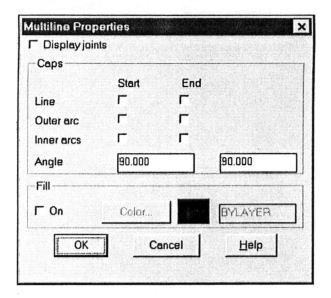

Examples shown below.

DISPLAY JOINTS **LINE** **OUTER ARC**

INNER ARC **ANGLE** **FILL**

LOADING A MULTILINE STYLE

The following instructions will guide you through how to LOAD an existing multiline style from one drawing to another.

1. Select the Multiline Style command using one of the following commands:

    TYPE = ML
    PULLDOWN = FORMAT / MULTILINE STYLE
    TOOLBAR = NONE

The following dialog box will appear.

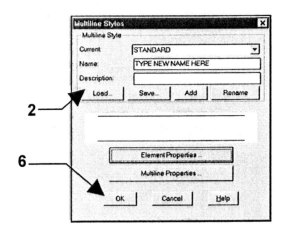

2. Select the LOAD button.

3. Select the FILE button.

4. Find the Multiline Style that you want to Load and select OPEN.

5. Select the OK button.

6. The loaded Mutiline style should be listed in the name box. It will also be the current style. Select the OK button.

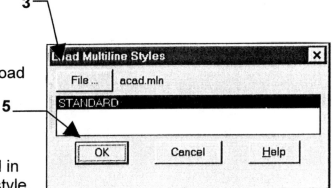

NOTE: If you insert a drawing file and it has a multiline style with the same name as an existing multiline file in the existing drawing on the screen, AutoCAD will change the incoming multiline style name by adding a "1" to the name.
Example: Walls will become Walls1.

EDITING MULTILINES

Multilines cannot be edited in the usual manner. Multiline has an editing feature all of its own. This makes multilines very powerful but somewhat difficult to use. You need to practice with the editing features to achieve the maximum results.

Select the Multiline Editing command using one of the following:

> **TYPE = Mledit**
> **PULLDOWN = MODIFY / OBJECT / MULTILINE**
> **TOOLBAR = MODIFY II**

The following dialog will appear.

CROSS
This column trims the *inner* intersecting lines in 3 different ways. (The first Multiline selected is trimmed.)

CORNER JOINT
This will trim intersecting lines to a right angle. Select the *inner* lines.

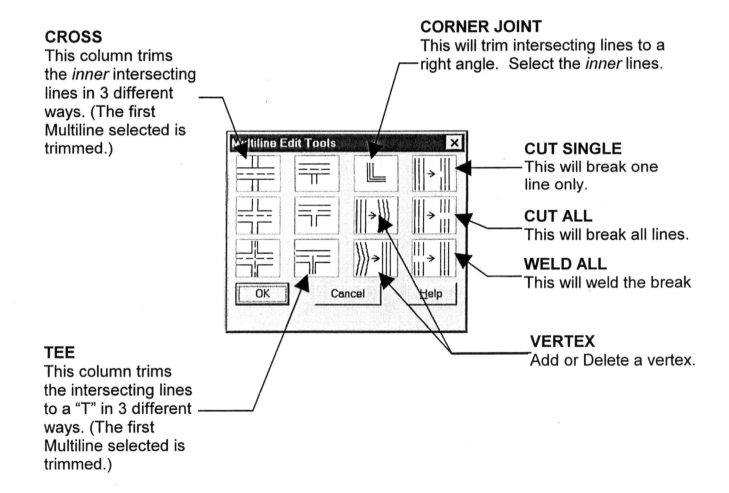

CUT SINGLE
This will break one line only.

CUT ALL
This will break all lines.

WELD ALL
This will weld the break

TEE
This column trims the intersecting lines to a "T" in 3 different ways. (The first Multiline selected is trimmed.)

VERTEX
Add or Delete a vertex.

DOUBLE LINES (LT users only)

The Double Line command allows you to draw two parrallel lines at the same time. This is a very easy and useful command. You specify the distance between the two lines (width), specify whether the two lines are open or closed at the ends (Caps) and a few other options explained below.

Note: Even though you draw both lines at the same time, each line is a separate entity and can be edited individually.

Walls on a floorplan would be an example of an architectural application. The thickness of a metal plate would be an example of a mechanical application.

1. Select the Double Line command using one of the following:
    TYPE = DLINE
    PULLDOWN = DRAW / DOUBLE LINE
    TOOLBAR = DRAW 🖉

    Command: _dline

2. Break/Caps/Dragline/Offset/Snap/Undo/Width/<start point>: *place the first endpoint or select an option*
3. Arc/Break/CAps/CLose/Dragline/Snap/Undo/Width/<next point>: *place the second endpt*
4. Arc/Break/CAps/CLose/Dragline/Snap/Undo/Width/<next point>: *place the next endpt or <enter> to stop.*

OPTIONS

BREAK Creates a gap where two double lines intersect. You can turn this ON or OFF. The default is ON.

ON   **OFF**

CAPS To set the ends of the double lines to be open or closed. The options are: **Both** – the start and the end, **Endpoint** – the end only, **Start point** – the first endpoint only, **None** – both ends open, **Auto** – close all end that are not joined with another double line.

Both   **End**   **Start**   **None**

DRAGLINE Basically this means "what is your cursor hanging on to?" Because there are two lines, you have to decide which line is the cursor attached to or do you want the cursor in the center of the two lines.   Your choices:

Right   **Center**   **Left**

OFFSET Allows you to start a double line at a specified distance and direction from a base point. First you are asked to locate the base point, next click in the direction you want the double line to start from that base point. Then the distance from the base point.

OPTIONS CONTINUED

SNAP Automatically snaps the new double line to an object close to the new endpoint of the double line. The default is ON. If you have Break and Snap ON, the new double line will automatically attach the new double line to the object and break a gap, in the object, between the double lines. Also you have the option of determining the **size** of the snapping area around the crosshairs. This area is called the *pixel search area*.

UNDO You undo the last double line segment. Must be done before you close or end the command.

WIDTH The distance between the two double lines.

DRAWING DOUBLE ARCS

1. Select the Double line command.
2. Place the first endpoint.
3. Arc/Break/CAps/CLose/Dragline/Snap/Undo/Width/<next point>: **A** <enter>
4. Break/CAps/CEnter/CLose/Dragline/Endpoint/Line/Snap/Undo/Width/<second point>:
    Place the second point for the arc's circumference or CE for the Arc's center point
5. Endpoint: ***Pick the second endpoint***

LAYER STATES

Layer States is a good organizational tool which can save you time and effort.

You already know that you can turn the layers on/off, freeze/thaw and lock/unlock. With **Layer States** you can save these settings as a "Layer State".

For example, let's say you have a floor plan drawing with walls, plumbing and electrical layers. From this drawing you would like to print 3 different drawings: First with walls only, second with walls and plumbing only and third with walls and electrical only. You can accomplish this by setting the freeze / thaw settings as shown below.

	Walls	**Plumbing**	**Electrical**
First print	Thaw	Freeze	Freeze
Second print	Thaw	Thaw	Freeze
Third print	Thaw	Freeze	Thaw

Now here is where the "Layer States" are helpful. You can set the freeze / thaw settings and save them as a "Layer State" with a Layer State "Name". You can "restore, edit or rename" the layer settings.

How to create NAMED LAYER STATES

1. Select the Layer command using one of the following:

   TYPE = LA
   PULLDOWN = FORMAT / LAYER
   TOOLBAR = OBJECT PROPERTIES

2. The dialog box shown, below left, will appear
3. Select the freeze / thaw settings, such as the "first print" above.
4. Select the "Save State" button.

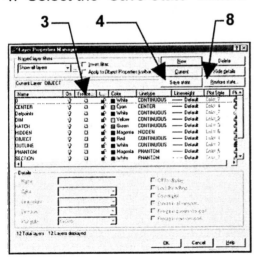

5. Enter the name in the **New layer state name** box. Select the <u>Layer states</u> and <u>properties</u> to be effected.

6. Select **OK**

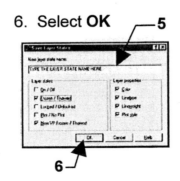

7. Repeat 3 thru 6 to create other layer states.

8. To **Restore** the saved layer states, select **Restore states**.

9. Select the name.

10. Select **Restore** button.

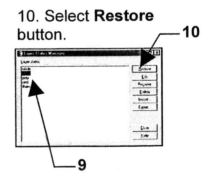

To practice this feature, do exercise 5D.

EXERCISE 5A
Create a MULTILINE STYLE

1. Open **My Decimal Setup**
2. Select Layout tab **11 X 17 (1 to 1)**
3. Create the Multiline Style shown below. Use the instructions "Creating a Multiline Style" on page 5-4.
4. NAME = **5A**
5. Save this drawing as **EX-5A.**

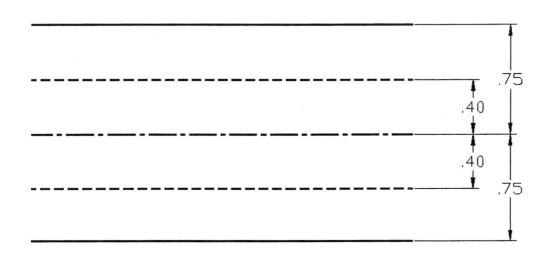

EXERCISE 5B
Drawing a MULTILINE

1. Open **My Decimal Setup**
2. Select Layout tab **11 X17 (1 to 1)**
3. Using the default multiline style **STANDARD**, draw the two objects below.
4. You must select the correct justification for each object. Pay close attention to where the dimensions are.
5. Save this drawing as **EX-5B.**

Set Justification to Bottom
START HERE
Location: 2, 8
draw Clock Wise.

Set Justification to Top
START HERE
Location: 8, 9
draw Clock Wise.

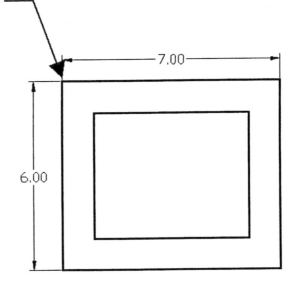

EXERCISE 5C
TRIMMING A MULTILINE

1. Open **My Decimal Setup**
2. Select Layout tab **11 X 17 (1 to 1)**
3. Using the multiline 5A, draw the 7 sets of crossing multilines below. Justification is not important.
4. Using **MODIFY / MULTILINE,** edit the multilines.
5. Save this drawing as **EX-5C.**

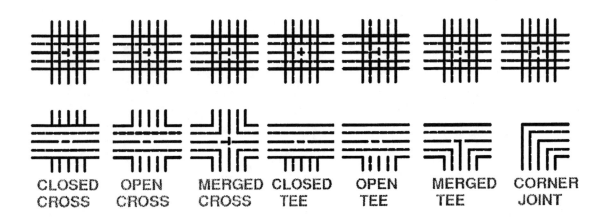

| CLOSED CROSS | OPEN CROSS | MERGED CROSS | CLOSED TEE | OPEN TEE | MERGED TEE | CORNER JOINT |

EXERCISE 5D
Create and Restore LAYER STATES

A. Open **My Decimal Setup**

B. Draw the 3 objects shown below on the following layers: (size is not important)
Circle = Object layer   Rectangle = Hatch layer   Polygon = Dim layer

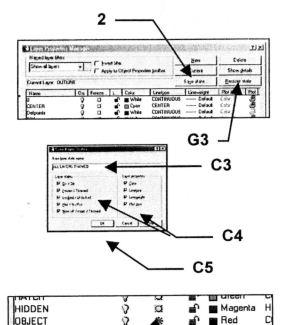

C. **Now you are going to create a layer state with "All Layers Thawed".**
1. Select **Format / Layer**
2. Select the **Save State...** button.
3. Enter layer state name "**All layers thawed**"
4. Select all of the States & Properties.
(check all boxes)
5. Select the **OK** button.

D. **Create a Layer State with only the Object layer frozen.**
1. **Freeze** the Object layer.
2. Select the **Save State...** button
3. Enter layer state name "**Circle Frozen**" and select **OK**.

E. **Create a Layer State with the Hatch layer Frozen.**
1. **Thaw** the **Object** Layer and **Freeze** the **Hatch** layer.
2. Select the **Save State...** button
3. Enter layer state name "**Rectangle Frozen**" and select **OK**

F. **Create a Layer State with Dim Layer Frozen.**
1. **Thaw** the **Hatch** layer and **Freeze** the Dim layer.
2. Select the **Save State...** button.
3. Enter layer state name "**Polygon Frozen**" and select **OK**.

G. You now have 3 layer states, let's see how they work.

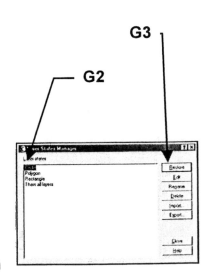

1. Select **Format / Layer** then select **Restore State...** button.
2. Select **Circle** from the list of Layer States.
3. Select the **Restore** button.
4. *Look at your list of layers. Is the Object layer Frozen? Now look at the drawing. Is the Circle missing?*
5. Now try the Rectangle and/or the Polygon Layer States. Each time look at the list of layers and then the drawing.
6. Save this drawing as: **EX-5D**. Do not plot.

EXERCISE 5E
CREATE A FLOOR PLAN USING MULTILINE

A. Open **My Feet-Inches Setup**

B. Select the **Qtr Equals Foot** tab.

C. Create 4 new MULTILINE STYLES.

1. Name = WALLS6	1. Name = WALLS4	1. Name = WINDOW	1. Name = Door
2. Offset = 3 and -3	2. Offset = 2 and -2	2. Offset = 3 & -3	2. Offset = 15 & -15
3. Color = Bylayer	3. Color = Bylayer	3. Color = Bylayer	3. Color = Bylayer
4. Linetype = Bylayer	4. Linetype = Bylayer	4. Linetype=Bylayer	4.Linetype = Bylayer

D. Draw the Floor Plan (refer to the drawing on page 5-17.)
   1. Use "WALLS6" for the exterior walls
   2. Use "WALLS4" for the interior walls
   3. Use Layer Walls
   4. Don't forget to set the JUSTIFICATION to BOTTOM and draw CW.
   5. Leave spaces for the Windows and Doors OR use Modify / Multiline / Cut all.

E. DOORS
   1. Use Layer DOORS
   2. Door Size = 30" wide X 2" thick (4" space behind door)

F. FURNITURE
   1. Use Layer FURNITURE
   2. Place approximately as shown.
   3. Sizes:

Desk = 3' x 5' (3' x 15" return)	Credenza's = 18" x 5'	Chairs = 2' Sq.
Sofa = 3' x 6'	Lamp Table = 2' Sq.	
File Cabinets = 15" x 24"	Copier = 2' x 3'	

G. ELECTRICAL
   1. Use Layer ELECTRICAL for switches and fixtures.
   2. Use Layer **WIRING** for the wire from switches to fixtures.
   3. Sizes:

Wall Outlet = 8" dia.	Overhead Lights = 16" dia.	Switches = Text (ht = 6")

H. AREA TITLES (OFFICE, LOBBY AND RECEPTION)
   1. Use Layer = Text Heavy   Text Style = Arch Text   Height = 9" (3/16 x 48 DSF)

I. Save as: Ex-5E

J. Plot all 4 drawings listed below using "Layer States":
   1. Complete drawing   (all layers thawed)
   2. Floor Plan only   (Freeze layers: Furniture, Electrical and Wiring)
   3. Floor Plan & Electrical   (Thaw layers: Electrical and Wiring)
   4. Floor Plan & Furniture   (Freeze layers: Elect. & Wiring.   Thaw layer: Furniture)

EXERCISE 5F
CREATE A FLOOR PLAN USING DOUBLE LINE

A. Open **My Feet-Inches Setup**

B. Select the **Qtr Equals Foot** tab.

C. Draw the Floor Plan (refer to the drawing on page 5-17.)
1. The exterior walls are 6" thick
2. The interior walls are 4" thick
3. Use Layer Walls
4. Don't forget to set the Dragline.
5. Leave spaces for the Windows and Doors OR use trim or break.

E. DOORS
1. Use Layer DOORS
2. Door Size = 30" wide X 2" thick (4" space behind door)

F. FURNITURE
1. Use Layer FURNITURE
2. Place approximately as shown.
3. Sizes:

Desk = 3' x 5' (3' x 15" return)	Credenza's = 18" x 5'	Chairs = 2' Sq.
Sofa = 3' x 6'	Lamp Table = 2' Sq.	
File Cabinets = 15" x 24"	Copier = 2' x 3'	

G. ELECTRICAL
1. Use Layer ELECTRICAL for switches and fixtures.
2. Use Layer **WIRING** for the wire from switches to fixtures.
3. Sizes:

Wall Outlet = 8" dia.   Overhead Lights = 16" dia.   Switches = Text (ht = 6")

H. AREA TITLES (OFFICE, LOBBY AND RECEPTION)
1. Use Layer = Text Heavy   Text Style = Arch Text   Height = 9" (3/16 x 48)

I. Save as: Ex-5F

J. **Plot all 4 drawings listed below**:
1. Complete drawing   (all layers thawed)
2. Floor Plan only   (Freeze layers: Furniture, Electrical and Wiring)
3. Floor Plan & Electrical   (Thaw layers: Electrical and Wiring)
4. Floor Plan & Furniture   (Freeze layers: Elect. & Wiring.   Thaw layer: Furniture)

EXERCISE 5E OR F

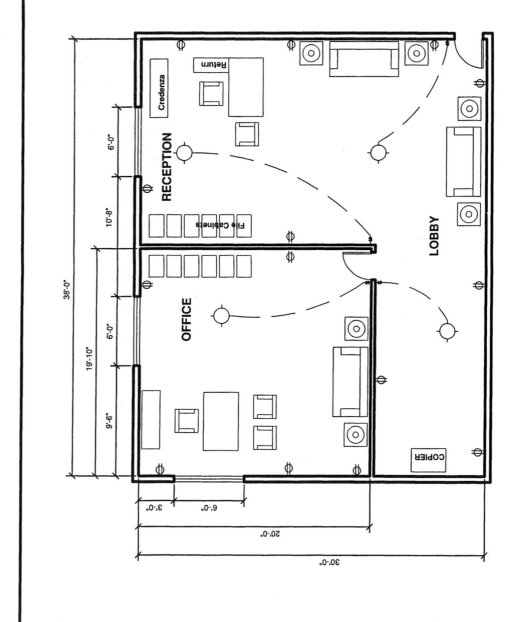

NOTES:

LEARNING OBJECTIVES

After completing this lesson, you will be able to:

1. Create an Isometric object.
2. Create an Ellipse on an Isometric plane.

LESSON 6

ISOMETRIC DRAWINGS

An ISOMETRIC drawing is a pictorial drawing. It is primarily used to aid in visualizing an object. There are 3 faces shown in one view. They are: Top, Right and Left. AutoCAD provides you with some help constructing an isometric drawing but you will be doing most of the work. The following are instructions on the isometric aids AutoCAD provides. **Note: This is not 3D. Refer the "Introduction to 3D" section in this workbook.**

ISOMETRIC SNAP and GRID

1. Select **TOOLS / DRAFTING SETTINGS** or right click on the Grid button then select **Settings** from the short cut menu.

   The following dialog box should appear.

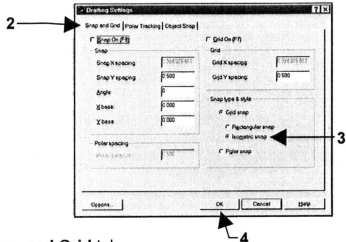

2. Select the **Snap and Grid** tab.
3. Select "**Isometric snap**"
4. Select **OK**

NOTE: The Grid pattern and Cursor have changed. The grids are on a 30 degree angle. The cursor's horizontal crosshair is now on a 30 degree angle also. This indicates that the LEFT ISOPLANE is displayed.

ISOPLANES

There are 3 isoplanes, Left, Top and Right. You can toggle to each one by pressing the **F5** key. When drawing the left side of an object you should display the Left Isoplane. When drawing the Top of an object you should display the Top Isoplane. When drawing the Right side of an object you should display the Right Isoplane.

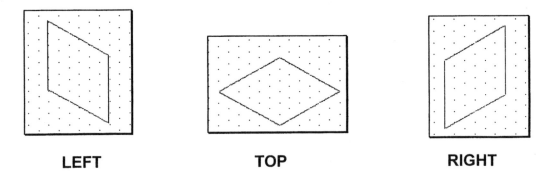

| LEFT | TOP | RIGHT |

5. **Grid, Snap and Ortho should be ON**.

6. Now try drawing the 3 x 3 x 3 cube below using "Line".
   a. Start with the Left Isoplane. (You may use **DIRECT DISTANCE ENTRY**)
   b. Next change to the Top Isoplane (Press F5 once) and draw the top.
   c. Next change to the Right Isoplane (Press F5 once) and draw the Right side.

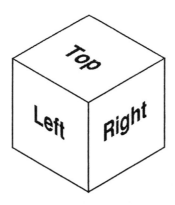

ISOMETRIC ELLIPSE

An isometric ellipse is a very helpful option when drawing an isometric drawing. AutoCAD calls an isometric ellipse an ISOCIRCLE. This option is located in the Ellipse command. ***IMPORTANT: the ISOCIRCLE option can only be selected when you have the Isometric Grid and Snap ON. This option will not appear if <u>Isometric Grid and Snap</u> is turned OFF.***

1. Change to the Left Isoplane. (Isometric Grid and Snap must be on.)
2. Select the **ELLIPSE** command. ***(Note: Use the Pull-down Menu, select "Axis,End" method. <u>Do not select "Center"</u>)***

   Command: _ellipse
   Specify axis endpoint of ellipse or [Arc/Center/Isocircle]: ***type "I" <enter>***
   Specify center of isocircle: ***specify the center location for the new isocircle***
   Specify radius of isocircle or [Diameter]: ***type the radius <enter>***

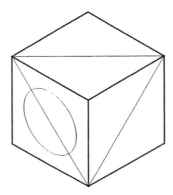

3. Now try drawing an Isocircle on the top and right, using the method above. Don't forget to change the "Isoplane" depending on where you draw the Isocircle.

EXERCISE 6A
ISOMETRIC ASSEMBLY

1. OPEN "MY DECIMAL SETUP"
2. SELECT THE "24-18-FULL" TAB.
3. CHANGE TO "MODEL SPACE"
4. MAKE SURE YOUR VIEWPORT SCALE IS SET TO 1:1 AND LOCKED.
5. SET THE "SNAP TYPE & STYLE" TO ISOMETRIC SNAP (Refer to page 6-2 if necessary)
6. CHANGE TO ISOPLANE "TOP".
7. DRAW THE OBJECTS BELOW. DO NOT DIMENSION.
   (DIMENSIONING WILL BE DISCUSSED IN LESSON 7.)
8. SAVE AS: EX-6A.

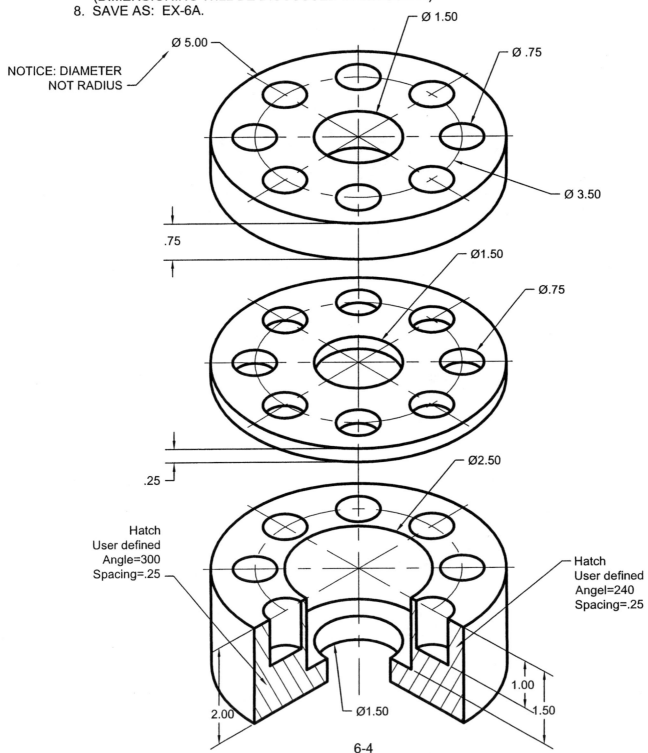

NOTICE: DIAMETER NOT RADIUS

Ø 5.00

Ø 1.50

Ø .75

Ø 3.50

.75

Ø1.50

Ø.75

.25

Ø2.50

Hatch
User defined
Angle=300
Spacing=.25

Hatch
User defined
Angel=240
Spacing=.25

1.00

1.50

2.00

Ø1.50

6-4

EXERCISE 6B

ISOMETRIC OBJECT

1. OPEN "MY DECIMAL SETUP"
2. SELECT THE "24-18-FULL" TAB.
3. CHANGE TO "MODEL SPACE"
4. MAKE SURE VIEWPORT SCALE IS SET TO TO 1:1 AND LOCKED.
5. SET THE "SNAP TYPE & STYLE" TO ISOMETRIC SNAP (Refer to page 6-2 if necessary)
6. CHANGE THE ISOPLANE WITH F5
7. DRAW THE OBJECTS BELOW. DO NOT DIMENSION.
   (DIMENSIONING WILL BE DISCUSSED IN LESSON 7.)
8. SAVE AS: EX-6B

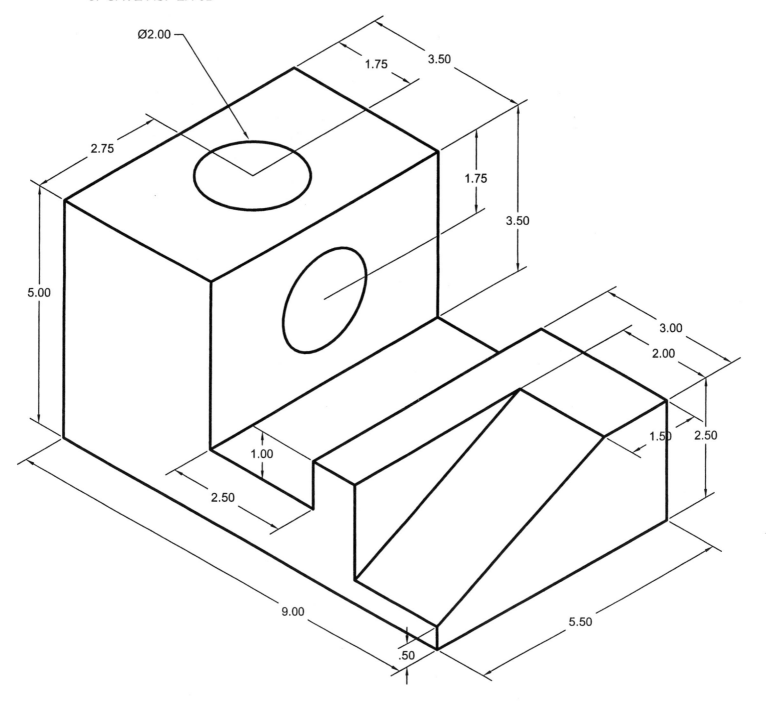

EXERCISE 6C
ABSTRACT HOUSE

1. Open "My Feet-Inches Setup"
2. Select the "Qtr Equals Foot" tab.
3. Change to Model Space
4. Make sure your viewport scale is 1/4" = 1' and Locked.
5. Draw the abstract house below. Do not dimension.
6. Save as Ex-6C.

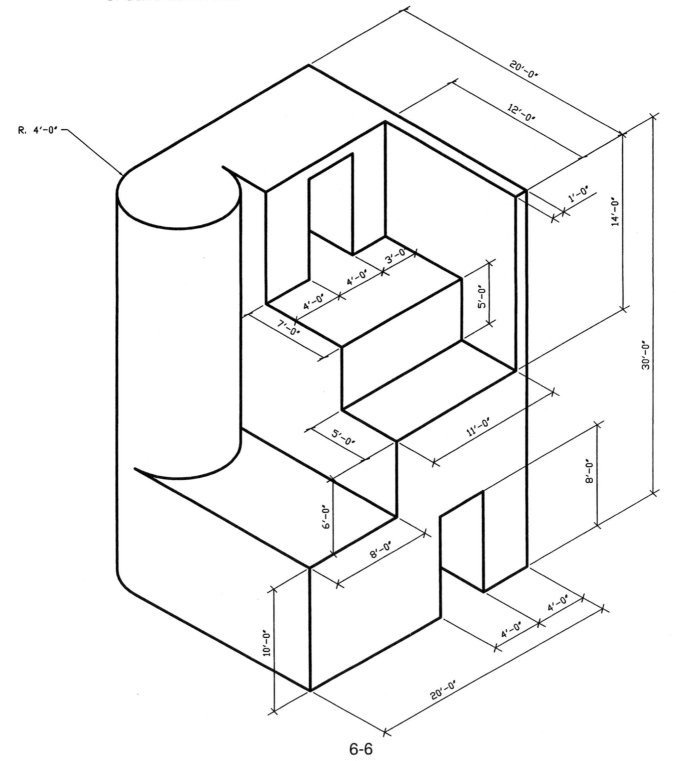

LEARNING OBJECTIVES

After completing this lesson, you will be able to:

1. Cut and Paste from one document to another.
2. Dimension an Isometric object.
3. Use Grips
4. Create Isometric Text.

LESSON 7

COPY, CUT and PASTE

What is COPY, CUT and PASTE?

AutoCAD allows you to select objects in a drawing **(COPY)** and then copy those objects to another drawing **(PASTE)**. Also, when you select the objects, you can make the selected objects disappear in the original drawing **(CUT)** and then copy **(PASTE)** them to another drawing.

You can also copy, cut and paste between AutoCAD and other documents such as Microsoft Word, Excel or any other Windows application.

Copy, Cut and Paste between AutoCAD drawings. (Simplified)

1. Open two AutoCAD drawings and Tile Vertically. (Example below)
2. Activate the drawing that contains the objects you want to copy. (Click in the drawing area)
3. Select the **COPY** or **CUT** command using one of the following:

TYPE = Copyclip	**TYPE = Cutclip**	**TYPE = Pasteclip**
PULLDOWN = EDIT / COPY	**PULLDOWN = EDIT / CUT**	**PULLDOWN = EDIT/PASTE**
TOOLBAR = STANDARD	**TOOLBAR = STANDARD**	**TOOLBAR = STANDARD**

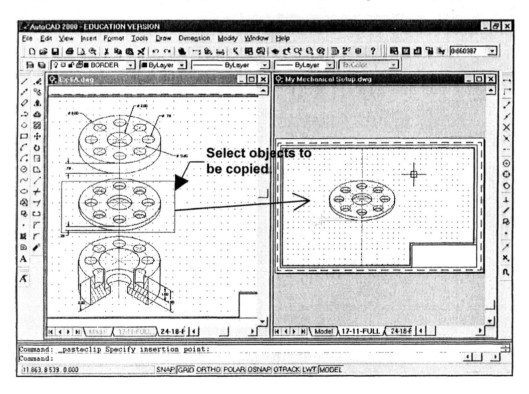

4. Select the objects to copy then <enter>. (Use any AutoCAD selection methods) If you selected **CUT**, the objects selected will disappear.
5. Activate the drawing in which the objects will be pasted.
6. Select the **PASTE** command.
7. Place the objects with the cursor by dragging the objects to the correct location and click.

Note: In the example above the base point for insertion was not designated by you. If you wish to designate a specific basepoint see **COPY with a Base Point** on the next page.

Where do the objects go before pasting?

Copies of the objects are temporarily stored on the Windows Clipboard. The Clipboard retains only the most recent information that was copied. When another object is copied, the newly copied object replace the last objects stored within the Clipboard's memory.

ADDITIONAL OPTIONS

COPY with a BASE POINT or COPYBASE
This option is the same as copy except it allows you to specify a base point for the objects.

COPY LINK
Copies **all** objects currently on the screen and automatically selects the ORIGIN as the basepoint. You will not be prompted to select objects.

PASTE as BLOCK or PASTEBLOCK
Copy objects into the same or different drawing as a BLOCK.

PASTE to ORIGINAL COORDINATES or PASTEORIG
Pastes objects into the new drawing at the same coordinate position as the original drawing. You will not be prompted for an insertion point.

PASTE SPECIAL or PASTESPEC
Use when pasting from other applications into AutoCAD.

Copy, Cut and Paste between other applications and AutoCAD. (Simplified)

*There are many options, within AutoCAD, for copy, cut and paste between other applications and AutoCAD. But the following describes the most commonly used option, "How to copy and paste text from a **word processing** program to an **AutoCAD drawing**".*

1. Open the Text document.
2. Highlight the text that you wish to copy.
3. Select the COPY command.
4. Open the AutoCAD drawing.
5. Select the PASTE command.

The following dialog box <u>will not</u> appear if you are using Windows, Notepad.

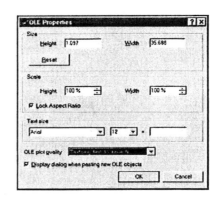

   a. **Size or Scale** changes the overall size of the pasted object. Changing one automatically changes the other.
   b. **Lock Aspect Ratio** box checked means the Height and Width will remain proportionate. Uncheck the box and the Height and Width can be set independently.
   c. **Text Size** and font can be changed here.
   d. Select OK button.

*After the text has been pasted, it can be changed in many ways. See **Editing OLE objects**.*

EDITING OLE OBJECTS

Microsoft developed a method called OBJECT LINKING and EMBEDDING or OLE. Text that has been pasted into a drawing is an OLE object and can be edited using any of AutoCAD's editing features.
In addition you may bring the text to the front or send it back behind other objects.

Bring to Front, send to Back, bring above object or send under object.

You may change the placement of the OLE object relative to other objects in the drawing. For example, you may want the pasted text to be in front of a drawing view. You would bring the text to the front and send the drawing view to the back.

1. Left click on the OLE object to activate it.
2. Select **TOOLS / DISPLAY ORDER**
3. Choose one of the following:
   Bring to Front
   Send to back
   Bring above object
   Send under object

DIMENSIONING AN ISOMETRIC DRAWING

Dimensioning an isometric drawing in AutoCAD is a two step process. First you dimension it with the dimension command **ALIGNED**. Then you adjust the angle of the extension line with the dimension command **OBLIQUE**.

STEP 1. Dimension the object using the dimension command **ALIGNED**.

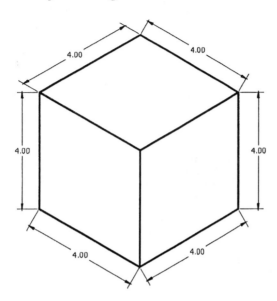

STEP 2. Adjust the angle of the extension lines using the dimension command **OBLIQUE.**
          (Do each dimension individually)
1. Select **DIMENSION / OBLIQUE** or *Type: **DIMEDIT** and select **O** for Oblique* or
2. Select the dimension you wish to adjust. (click on it) <enter>
3. Type the Obliquing angle for the **extension line**. (30, 150, 210 or 330)

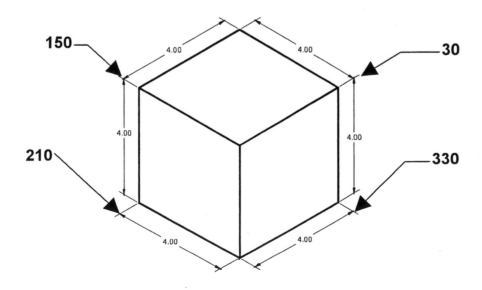

7-5

GRIPS

Grips are little boxes that appear if you select an object when no command is in use. Grips must be enabled by typing "grips" <enter> then 1 <enter> on the command line or selecting the "enable grips" box in the OPTIONS / SELECTION dialog box.

Grips can be used to quickly edit objects. You can move, copy, stretch, mirror, rotate, and scale objects using grips.

The following is an overview on how to use three of the most frequently used options. Grips have many more options and if you like the example below, you should research them further in the AutoCAD help menu.

1. Select the object (no command can be in use while using grips)
2. Select one of the grips. It will turn solid. This indicates that it is "hot". Hot means that this grip is the basepoint.
3. Select the edit mode (move, copy, stretch) from the command line. You can cycle through these modes with the SPACEBAR or ENTER keys or shortcut menu.
4. <u>After editing you must press the ESC key two times to deactivate the grips on that object.</u>

Selecting a grip:
When you select a grip it becomes "HOT".

| Hot grip |

Moving an object:
1. Select the object.
2. Select the grip in the middle of the object.
3. Move the cursor to the new location and press the left mouse button.

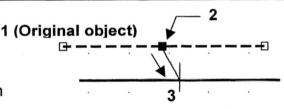

| Move |

Copying an object:
1. Select the object.
2. Select the middle grip.
3. Right click and select Copy from the short cut menu
4. Select the new location for the copy(s).
Note: Grips will allow you to make multiple copies. Press the ESC key to stop.

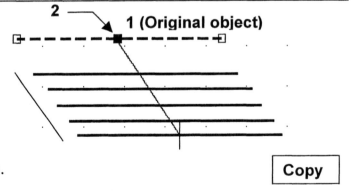

| Copy |

Stretch an object:
1. Select the object
2. Select an end grip.
3. Move the cursor to stretch the object.
4. Press left mouse button to place.

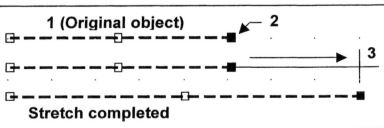

Stretch completed

| Stretch |

ISOMETRIC TEXT

AutoCAD doesn't really have an Isometric Text. But by using both **Rotation** and **Oblique angle**, we can make text appear to be laying on the surface of an Isometric object.

1. First create two new text styles. One with an <u>oblique angle</u> of **30** and the other with an <u>oblique angle</u> of **minus 30**.

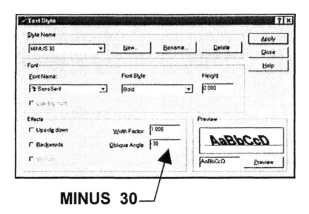

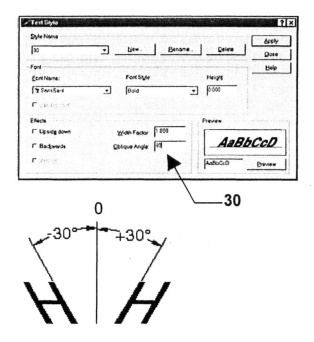

2. Next select the appropriate text style and rotation as follows:

   a. Select the text style with an obliquing angle of 30 or minus 30.
   b. Select **DRAW / TEXT / SINGLE LINE**
   c. Place the START POINT or Justify.
   d. Type the Height.
   e. Type the Rotation angle.
   f. Type the text.

This is an example of what you can achieve. The text appears to be on the top and side surfaces.

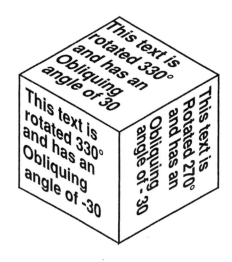

EXERCISE 7A

Copy and Paste between two drawings.

1. Confirm that "Single-drawing Compatibility mode" is OFF (Refer to page 1-2)
2. Open **My Decimal Setup** and select the **11 X 17 (1 to 1)** tab.
3. Also, open **Ex-6A**.
4. Tile Vertically. (Example below)
5. Activate the Ex-6A drawing (Click in the drawing area)
6. Select the **COPY** command using one of the following:

> **TYPE = Copyclip or CTRL + C**
> **PULLDOWN = EDIT / COPY**
> **TOOLBAR = STANDARD**

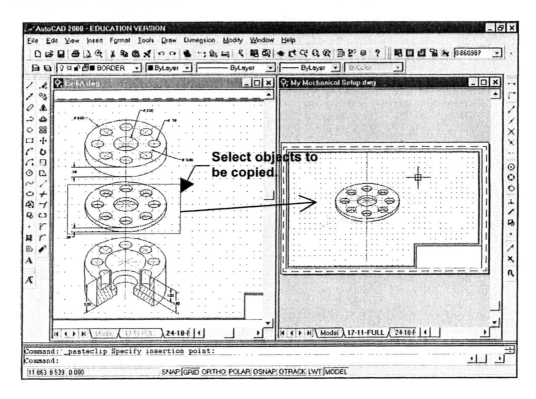

7. Select the objects to copy with a crossing window then <enter>.
8. Activate the **My Decimal Setup** drawing.
9. Select the **PASTE** command using one of the following methods.

> **TYPE = Pasteclip or CTRL + V**
> **PULLDOWN = EDIT / PASTE**
> **TOOLBAR = STANDARD**

10. Place the objects with the cursor in the drawing area and left click. (As shown above.)

11. **Do not save** this drawing. It is for practice only.

EXERCISE 7B
OBLIQUE DIMENSIONING

1. OPEN DRAWING 6B.
2. DIMENSION THE ISOMETRIC OBJECT. (Refer to page 7-5 if necessary)
3. USE "DIMENSION / ALIGNED" FIRST.
4. THEN "DIMENSION / OBLIQUE.
   (REMEMBER, THE OBLIQUE ANGLE IS THE DESIRED ANGLE FOR THE EXTENSION LINE.)
5. USE GRIPS TO MOVE THE DIMENSIONS IF NECESSARY.
6. SAVE AS: 7B

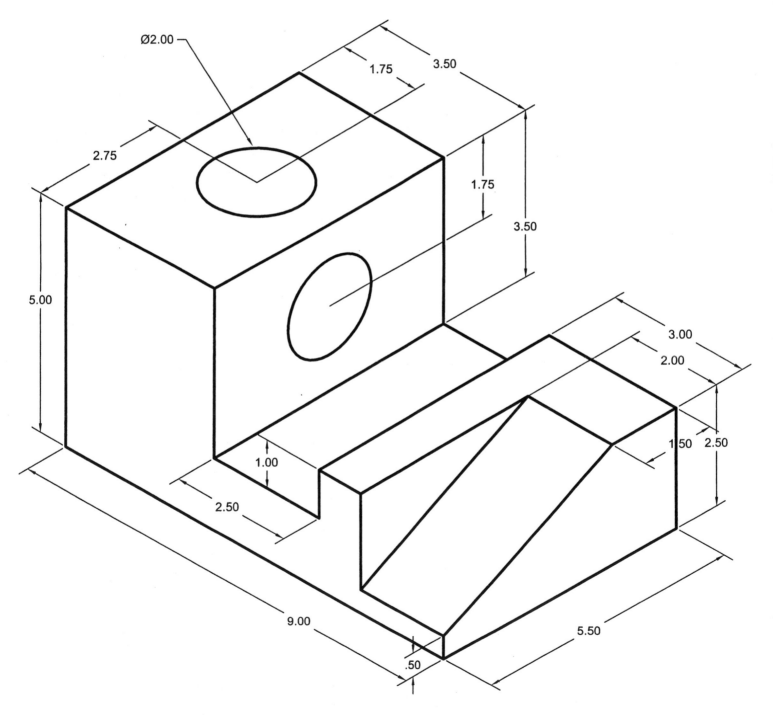

EXERCISE 7C
OBLIQUE DIMENSIONING

1. Open drawing 6C.
2. Dimension the abstract house.
3. Use "Dimension / Aligned" first. (dimension in paperspace)
4. Now use "Dimension / Oblique"
5. Use grips to move the dimensions if necessary.
6. Save as Ex-7C.

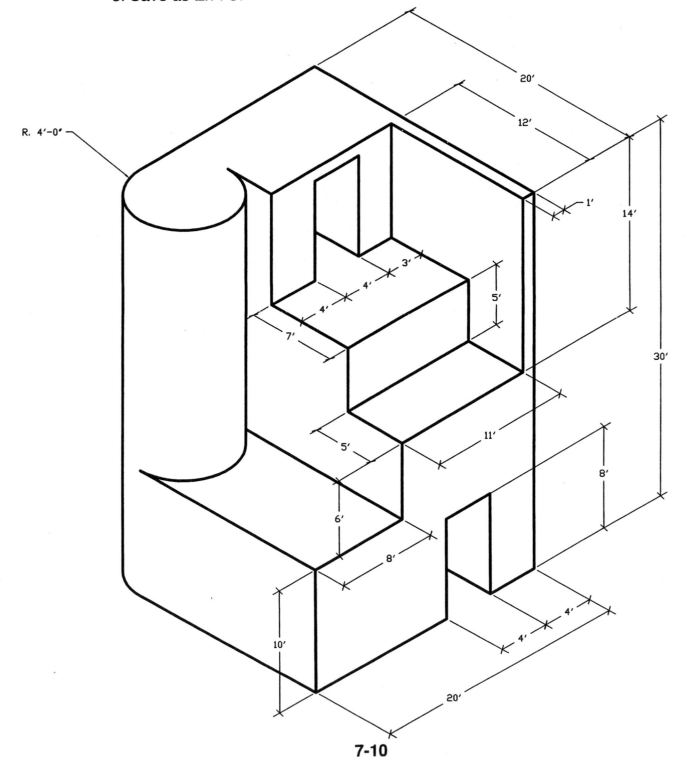

EXERCISE 7D

ISOMETRIC TEXT

1. Open My Decimal Setup.
2. Select the 11 X 17 (1 to 1) tab.
3. First create two new text styles. One with an oblique angle of 30 and the other with an oblique angle of minus 30.

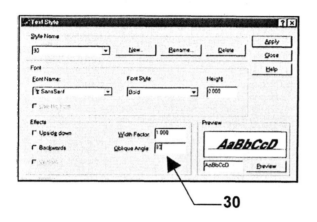

30

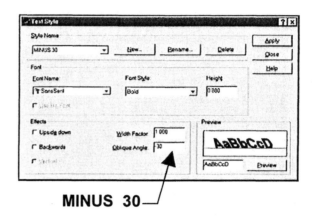

MINUS 30

4. Draw a 4 inch cube.

5. Next select the appropriate text style and rotation to create the text on the cube shown below.

   a. Select the text style with an oblique angle of 30 or minus 30.
   b. Select **DRAW / TEXT / SINGLE LINE**
   c. Place the START POINT approximately as shown.
   d. Type the Height = .25
   e. Type the Rotation angle.
   f. Type the text.

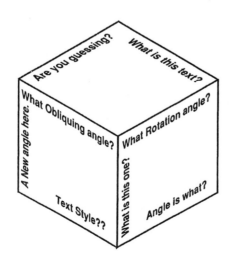

NOTES:

LEARNING OBJECTIVES

After completing this lesson, you will be able to:

1. Create a Block (Review)
2. Insert a Block (Review)
3. Assign and use Attributes

LESSON 8

Let's review "<u>How to create a block</u>"

You learned how to create Blocks, in Lesson 28, in the Exercise Workbook for Beginning AutoCAD 2004. But because Blocks are so useful, let's review "How to create a block" one more time. Then we will learn how to add "Attributes" to those blocks to make them even more useful.

CREATING A BLOCK

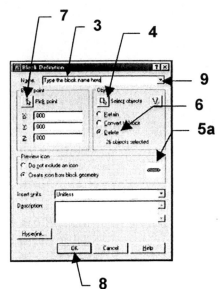

1. First draw the objects that will be converted into a Block. (Refer to page 8-3 "**Color and Linetype**")

2. Select the **BMAKE** command using one of the following:

   TYPE = B
   PULLDOWN = DRAW / BLOCK / MAKE
   TOOLBAR =DRAW

   (the dialog box, on the right, will appear)

3. Enter the New Block name in the **Name** box.

4. Select the **SELECT OBJECTS** button.
   The Block Definition box will disappear and you will return temporarily to the drawing.

5. Select the objects you want in the block, then press <enter>
   The Block Definition box will reappear and the objects you selected should be illustrated in the Preview Icon area.

6. Select **Delete** (Refer to 8-3 for definitions)

7. Select the **Pick Point** button. (Or you may type the X and Y coordinates)
   The Block Definition box will disappear again and you will return temporarily to the drawing.

   Select the location where you would like the insertion point for the Block. Later when you insert this block, it will appear on the screen attached to the cursor at this insertion point. Usually this point is the CENTER, MIDPOINT or ENDPOINT of an object.

8. Select the **OK** button.
   The objects will disappear but the new block is now stored in the drawing's block definition table.

9. To verify the creation of this Block, select the Block command, select the Name (▼). A list of all the blocks, in this drawing, will appear.

ADDITIONAL DEFINITIONS OF OPTIONS.

Retain
If this option is selected, the original objects will stay visible on the screen after the block has been created.

Convert to block
If this option is selected, the original objects will disappear after the block has been created, but will immediately reappear as a block. It happens so fast, you won't even notice the original objects disappeared.

Delete
If this option is selected, the original objects will disappear from the screen, after the block has been created.

Do not include an icon or Create icon from block geometry
These options determine whether a " preview thumbnail" sketch is created. This option is used with the "Design Center" to drag and drop the blocks into a drawing. The Design Center is an advanced option and is not discussed in this book.

Insert Units
You may define the units of measurement for the block. This option is used with the "Design Center" to drag and drop with Autoscaling. The Design Center is an advanced option and is not discussed in this book.

COLOR and LINETYPE

If a block is created on Layer 0:

When the block is inserted, it will assume the Properties, color, linetype etc. of the layer that is current at the time of insertion.
The block will also reside on the layer that was current at the time of insertion.
If the Block is then **Exploded**, the objects included in the block, will go back to their original color, linetype and layer.

If a block is created on Specific layers:

When the block is inserted, it will retain its own Properties, color, linetype etc. It **will not assume** the color and linetype of the layer that is current.
But the block **will reside** on the current layer at the time of insertion.
If the Block is then **Exploded**, the objects included in the block will go back to their original layer. The color and linetype remain the same.

Let's review "How to Insert a Block"

Now we need to review "How to Insert the Block" you just created.

1. Select the INSERT command using one of the following:
   TYPE = DDINSERT
   PULLDOWN = INSERT / BLOCK
   TOOLBAR =DRAW

The INSERT dialog box will appears.

2. Select the BLOCK name then select the OK button.
   a. If the block is in the drawing that is open on the screen:
      select the block from the drop down list shown above
   b. The Browse button can be used to find an entire drawing.

   This returns you to the drawing and the selected block should be attached to the cursor.

3. Select the location for the block by pressing the left mouse button or typing coordinates.

NOTE: If you want to scale or rotate the block before you actually place the block, press the right hand mouse button, and you may select an option from the short cut menu or select an option from the command line menu shown below.

Command: _insert
Specify insertion point or **[Scale/X/Y/Z/Rotate/PScale/PX/PY/PZ/PRotate]:**

You may also "preset" the insertion point, scale or rotation. This is discussed on Page 8-5.

PRESETTING THE "INSERTION POINT", "SCALE" OR "ROTATION"

You may want to specify the **Insertion point, Scale or Rotation** in the "INSERT" box instead of at the command line.

1. Remove the check mark from the **"SPECIFY ON-SCREEN"** box.
2. Fill in the appropriate information describe below:

Insertion point
Type the X and Y coordinates from the Origin. The Z is for 3D only.
The example below indicates the blocks insertion point will be 5 inches in the X direction and 3 inches in the Y direction, from the Origin.

Scale
You may scale the block proportionately by typing the scale factor in the X box and then check the Uniform Scale box.
If the block will be scaled non-proportionately, type the different scale factors in both X and Y boxes.

The example below indicates that the block will be scale proportionate at a factor of 2.

Rotation
Type the desired rotation angle relative to its current rotation angle.

The example below indicates the block will be rotated 45 degrees from it's originally created angle orientation.

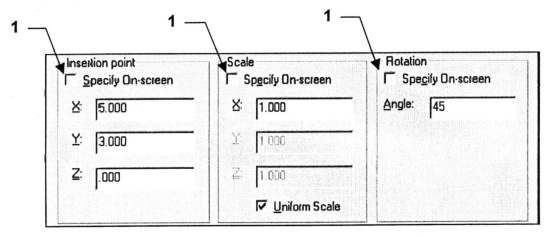

ATTRIBUTES

The **ATTRIBUTE** command allows you to add text data to a block. You define the attribute and attach it to a block. Every time you insert the block, the attributes are also inserted.

For example, if you had a block in the shape of a tree, you could assign information (attributes) about this tree, such as name, size, cost, etc. Each time you insert the block, AutoCAD will pause and prompt you for "What kind of tree is this?" You respond by entering the tree name. Then another prompt will appear, "What size is this tree? You respond by entering the size. Then another prompt will appear, "What is the cost"? You respond by entering the cost. The tree symbol will then appear in the drawing with the name, size and cost displayed.

Another example could be a title block. If you assign attributes to the text in the title block, when you insert the title block it will pause and prompt you for "What is the name of the drawing?" or "What is the drawing Number?". You respond by entering the information. The title block will then appear in the drawing with the information already filled in.

You will understand better after completing the following.

CREATING BLOCK ATTRIBUTES

1. Draw a rectangle 2 " X 1"

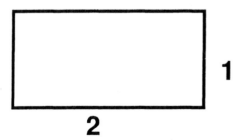

1

2

2. Select the **Define Attribute** command using one of the following:

   TYPE = ATTDEF
   PULLDOWN = DRAW / BLOCK / DEFINE ATTRIBUTES
   TOOLBAR = ATTRIBUTE

The following dialog box will appear.

Definitions on page 8-9

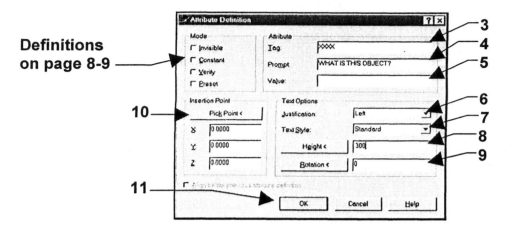

3. Enter the Attribute tag.
   A "Tag" is only a place saver. The "XXX's" will be replaced with your answer to the prompt when the block is inserted.

4. Enter the Attribute prompt.
   This is the prompt that will appear when AutoCAD pauses and prompts you for information. (You create the prompt)

5. Enter a value if necessary.
   The value will appear beside the prompt, in brackets, indicating what format the prompt is requesting. Example: What is the cost? <$0.00> The $0.00 inside the brackets is the "Value".

6. Select the Justification for the text.
   This will be used to place the Attribute text.

7. Select a Text Style.
   The Text Style selected will be used when displaying the Attribute.

8. Enter the text Height.
   This will be the height of the Attribute text.

9. Enter the Rotation.
   This will be the rotation angle of the Attribute text.

10. Select the Pick Point button.
    You will temporarily return to the drawing. Snap to the location where you would like the attribute to appear.

Snap to the corner

11. Select the OK button. Your rectangle should now look like this. *But we're not done yet.*

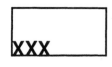

12. Select **DRAW / BLOCK / MAKE**. (Refer to page 8-2)
    Make a block of the rectangle and the attribute text.
    Name = abc
    Pick Point = upper right corner of rectangle
    Select Objects = Select rectangle and attribute text with a crossing window.

NOTE: Set the "ATTDIA" variable to 0 or 1

If the variable **ATTDIA** is set to **0**, the prompt will appear on the command line.

If the variable **ATTDIA** is set to **1**, the dialog box shown here will appear.

It is strictly personal preference.
It will not affect the input.
(Personally, I prefer the "dialog box")

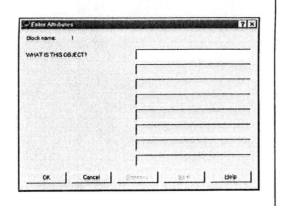

---

13. Select **INSERT / BLOCK** (Refer to page 8-4)
    a. Select block name **abc** and then **OK** button.
    b. Place the block somewhere on the screen and press the left mouse button.
    c. Type: **BOX <enter>**

14. Select **INSERT / BLOCK again.**
    a. Select block name **abc** and then **OK** button.
    b. Place the block somewhere on the screen and press the left mouse button.
    c. Type: **SQUARE <enter>**

15. Select **INSERT / BLOCK again.**
    a. Select block name **abc** and then **OK** button.
    b. Place the block somewhere on the screen and press the left mouse button.
    c. Type: **DESK <enter>**

You should now have 3 rectangles that look like the ones below. You inserted the same block but the text is different in each. Think how this could be useful in other applications.

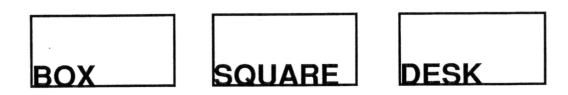

ATTRIBUTE MODES

Invisible
If this box is checked, the attribute will be invisible. You can make it visible later by typing the ATTDISP command and selecting ON.

Constant
The attribute stays constant. It never changes and you will not be prompted for the value when the block is inserted.

Verify
When you are prompted for input, after you type the input, you will be prompted to verify that input. This is a way to double check your input before it is entered on the screen. This option does not work when ATTDIA is set to 1 and the dialog box appears.

Preset
You will not be prompted for input. When the block is inserted, it will appear with the default value. You can edit it later with **ATTEDIT**. (See Editing Attributes page 9-2)

EXERCISE 8A

Assigning Attributes to a Block

The following exercise will instruct you to draw a Box, Assign Attributes, create a block of the Box with attributes, then insert the new block and answer the prompts when they appear on the screen.

1. Open **My Decimal Setup** and select the **11 X 17 (1 to 1)** tab.

2. Draw the Box shown on the next page.
   L = 3.75   W = 2.25   H = 3.00

3. Assign Attributes for **Length**
   a. Tag = L
   b. Prompt = What is the Length of the Box?
   c. Value = inches
   d. Justification = Middle
   e. Text Style = Standard
   f. Height = .250
   g. Rotation = 0

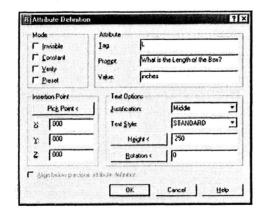

4. Assign Attributes for **Width**
   a. Tag = W
   b. Prompt = What is the Width of the Box?
   c. Value = inches
   d. Justification = Middle
   e. Text Style = Standard
   f. Height = .250
   g. Rotation = 0

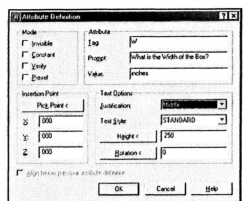

5. Assign Attributes for **Height**.
   a. Tag = H
   b. Prompt = What is Height of the Box?
   c. Value = inches
   d. Justification = Middle
   e. Text Style = Standard
   f. Height = .250
   g. Rotation = 0

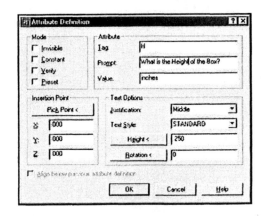

6. Assign Attributes for **Manufacturer**.
   a. Tag = MMMMM
   b. Prompt = Who is the Manufacturer?
   c. Value = (leave blank)
   d. Justification = Left
   e. Text Style = Standard
   f. Height = .250
   g. Rotation = 0

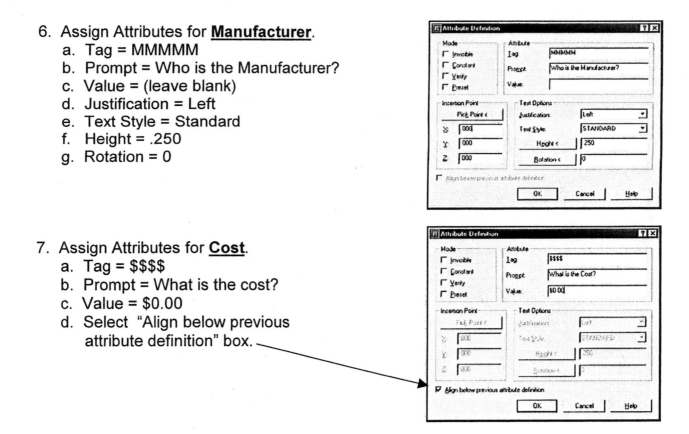

7. Assign Attributes for **Cost**.
   a. Tag = $$$$
   b. Prompt = What is the cost?
   c. Value = $0.00
   d. Select "Align below previous attribute definition" box.

Your drawing should look approximately like the example below.

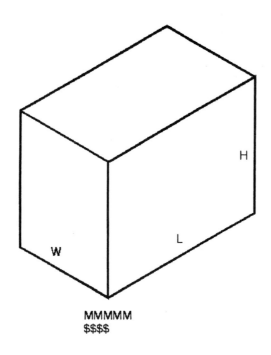

MMMMM
$$$$

8. When you have finished assigning all of the attributes, create a **BLOCK**. (Select the Box and the Attribute text.) (Refer to page 8-2 for instructions if necessary)

9. Now **Insert** the new block anywhere on the screen using
   a. **INSERT / BLOCK**.
   b. Browse to find the block
   c. Select OK

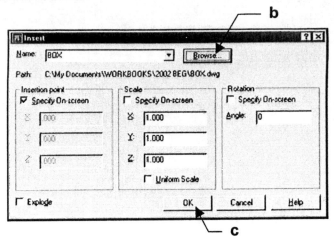

10. Type the answers in the box beside the attribute prompts as shown below.

Note: If the dialog box shown below does not appear, at the command line type ATTDIA <enter> and 1 <enter>.

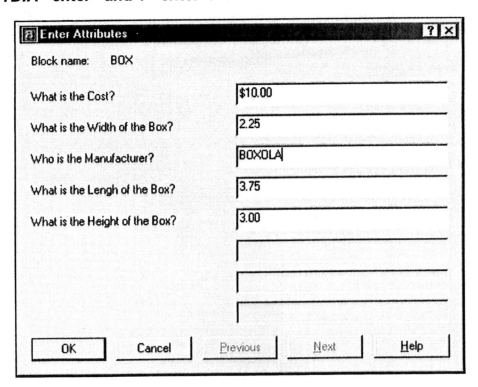

Notice that the prompts are not in any specific order. You can control the initial order of the prompts, when creating the block, by selecting the attribute text one by one with the cursor instead of using a window to select all objects.
In Lesson 9 you will learn how to rearrange the order after the block has been created.

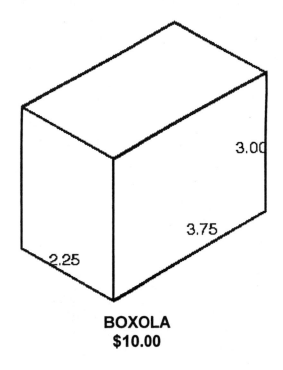

BOXOLA
$10.00

Does your box look like the example shown above?

11. Save as **EX-8A.**
12. **Do not plot.**

EXERCISE 8B

Assigning multiple Attributes to multiple Blocks

The following exercise will help you learn how to assign multiple attributes to 3 different objects. This drawing will be used in lesson 9 to extract the information and place it in a spreadsheet such as Excel.

A. Open **Ex-5E (or Ex-5F if you have AutoCAD LT)**

B. Assign the attributes to the <u>Sofa in the lobby</u>, the <u>Chair behind the office desk</u> and <u>one File Cabinet in the office</u>.

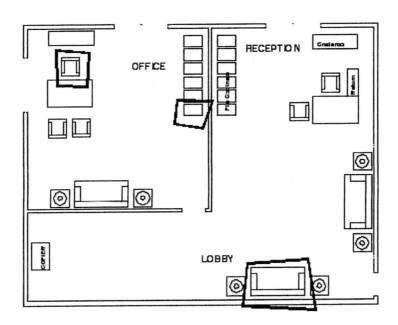

<u>*NOTE: The Attribute details for each object are on the next pages.*</u>

Before you start, check the ATTDIA setting. (Refer to 8-8)

C. LET'S START WITH THE SOFA

This is what the sofa should look like after you have completed the Attribute definitions shown below.

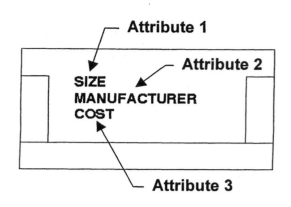

STEP 1

1. Select **Draw / Block / Define Attributes**
2. Fill in the boxes for Attribute 1.
3. Select the **Pick Point** button and place the Tag (Size) approximately as shown.
4. Repeat steps 2 and 3 for Attributes 2 and 3.

ATTRIBUTE 1

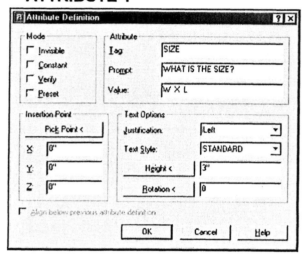

ATTRIBUTE 2

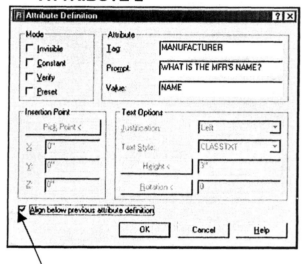

ATTRIBUTE 3

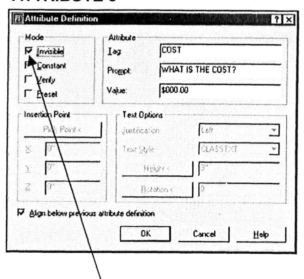

This option will align the new attribute directly below the previous and duplicate the Text options.

STEP 2

5. Now **create a block**.
    a. Select **Draw / Block / Make**
    b. Name = Sofa
    c. Select objects = select the sofa and the attribute text.
    d. Pick Point = Any corner of the sofa.

Note: the tag will not be invisible until you insert the block.

D. __NOW DO THE CHAIR__

This is what the chair should look like after you have completed the Attribute definitions shown below.

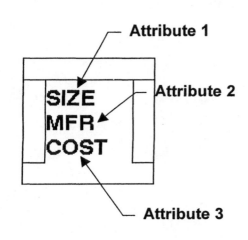

Attribute 1
Attribute 2
Attribute 3

__STEP 1__

1. Select __Draw / Block / Define Attributes__
2. Fill in the boxes for Attribute 1.
3. Select the __Pick Point__ button and place the Tag (Size) approximately as shown.
4. Repeat steps 2 and 3 for Attributes 2 and 3.

ATTRIBUTE 1

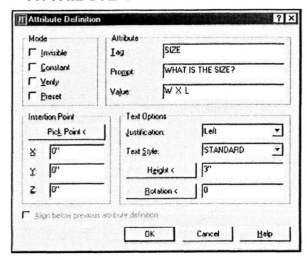

ATTRIBUTE 2

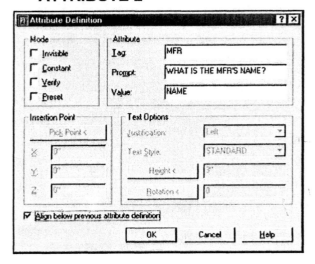

ATTRIBUTE 3

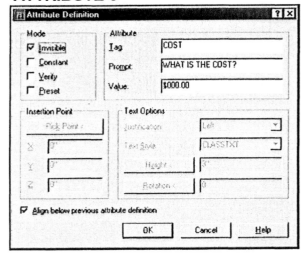

__STEP 2__

5. Now __create another block__.
   - a. Select __Draw / Block / Make__
   - b. Name = Chair
   - c. Select objects = select chair and the attribute text.
   - d. Pick Point = Any corner of the chair.

E. NOW DO THE FILE CABINET

This is what the file cabinet should look like after you have completed the Attribute definitions shown below.

Attribute 1
Attribute 2

SIZE
MFR
COST

Attribute 3

STEP 1

1. Select **Draw / Block / Define Attributes**
2. Fill in the boxes for Attribute 1.
3. Select the **Pick Point** button and place the Tag (Size) approximately as shown.
4. Repeat steps 2 and 3 for Attributes 2 and 3.

ATTRIBUTE 1

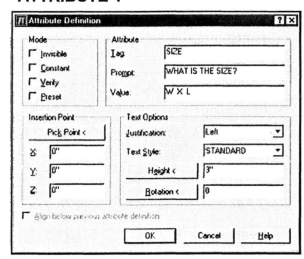

ATTRIBUTE 2

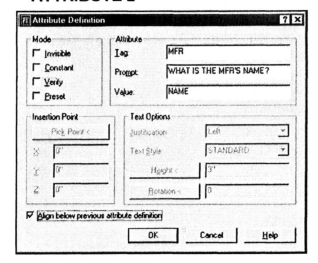

ATTRIBUTE 3

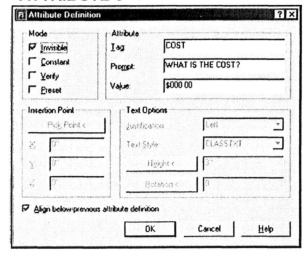

STEP 2

5. Now create a block.
   a. Select Draw / Block / Make
   b. Name = File cabinet
   c. Select objects = select the File cabinet and the attribute text.
   d. Pick Point = Any corner of the File cabinet.

Save this drawing at this point to be safe, so you don't lose anything.

F. ERASE all of the Sofas, Chairs and File Cabinets in the drawing.

G. INSERT the **Sofa block** into the same location as ONE of the sofas that you previously erased.
1. Select **INSERT / BLOCK**
2. Select the **SOFA** block from the drop down list.
3. Select **OK**
4. Answer the Attribute prompts:

      What is the SIZE?   3' X 6'
      Who is the Manufacturer? Sears
      What is the Cost? $500.00

5. Select **OK**.

Note: You can use the copy "mulitiple" command to place the remaining sofas.

Now your sofa should look approximately like the sofa below. *Notice the Cost is invisible.*

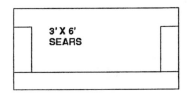

H. INSERT the **Chair block** into the same location as **ONE** of the chairs that you previously erased, as follows.

1. Select **INSERT / BLOCK**
2. Select the **CHAIR** block from the drop down list.
3. Select **OK**
4. Answer the Attribute prompts:

      What is the SIZE?   2' X 2'
      Who is the Manufacturer? LAZYBOY
      What is the Cost? $200.00

5. Select OK.

Note: You can use the copy "multiple" command to place the remaining chairs.

Now your Chair should look approximately like the chair below. *Notice the Cost is invisible.*

I. INSERT the **File cabinet block** into the same location as **ONE** of the File Cabinets that you previously erased as follows.

    1. Select **INSERT / BLOCK**
    2. Select the **File Cabinet** block from the drop down list.
    3. Select **OK**
    4. Answer the Attribute prompts:

        What is the SIZE?   15" X 24"
        Who is the Manufacturer? HON
        What is the Cost? $40.00

    5. Select **OK**.

Note: You can use the copy "multiple" command to place the remaining File Cabinets.

Now your File cabinet should look approximately like the File Cabinet below.
<u>*Notice the Cost is invisible.*</u>

```
┌────────────┐
│15" X 24    │
│HON         │
│            │
└────────────┘
```

<u>*Your drawing should look approximately like the example on the next page.*</u>

J. Save this drawing as: **EX-8B**

Do Not plot this drawing.

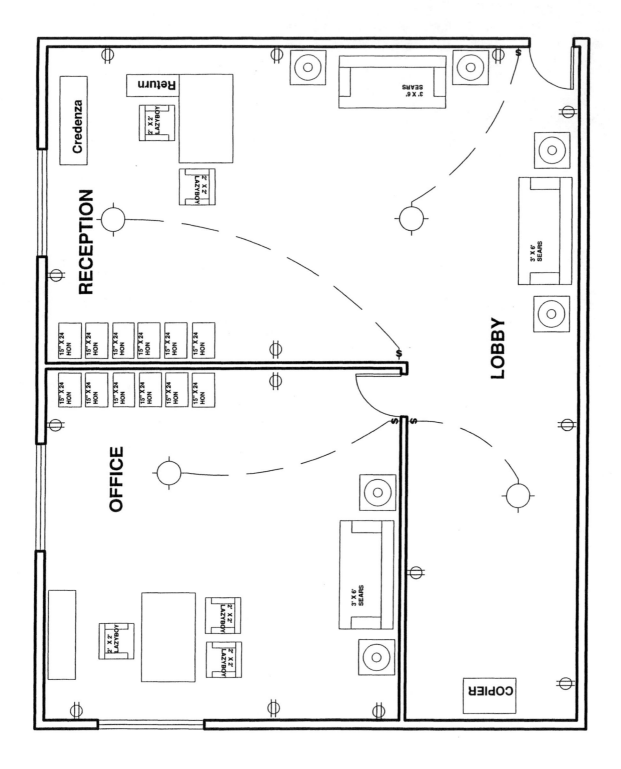

EXERCISE 8B

LEARNING OBJECTIVES

After completing this lesson, you will be able to:

1. Edit Attributes
2. Extract Attributes
3. Extract data to an Excel spreadsheet

LESSON 9

EDITING ATTRIBUTES

After a block with attributes has been inserted into a drawing you may want to edit it. AutoCAD has many ways to edit these attributes.

1. Select the **BLOCK ATTRIBUTE MANAGER** command using one of the following:

TYPE = BATTMAN
PULLDOWN = MODIFY / OBJECT / ATTRIBUTE / BLOCK ATTRIBUTE MANAGER
TOOLBAR = MODIFY II

2. Select the block that you want to edit.

The following dialog box will appear.

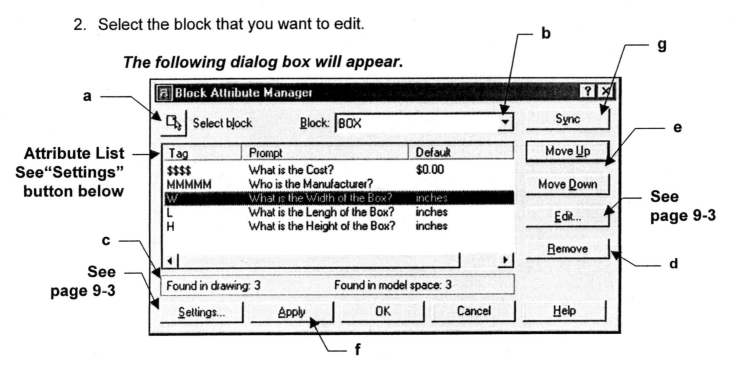

Attribute List
See"Settings"
button below

See
page 9-3

See
page 9-3

a. Select Block – Allows you to select another block – takes you back to the drawing so you can select another block to edit.

b. Block down arrow – Allows you to select another block –from a list.

c. Found – lists how many, of the selected block, were found in the drawing and how many are in the space you are currently in. (You must select the model "tab" or Layout "tab". It will not find model space attributes if you are not in model space.)

d. Remove - Allows you to remove an Attribute from a block. (See "Apply and Sync" below)

e. Move Up and Down - Allows you to put the prompts in the order you prefer.

f. Apply – After you have made all the changes, select the Apply button to update the attributes. (The Apply button will be gray if you have not made a change)

g. Sync – If you explode a block, make a change and redefine the block, Sync allows you to update all the previous blocks with the same name.

SETTINGS [Settings...]

The Settings dialog box controls which attributes are displayed. On the previous page TAG, PROMPT, DEFAULT and MODE are displayed. TAG values are always displayed.

a. Emphasize duplicate tags
If this option is ON, any duplicate tags will display RED.

b. Apply changes to existing ref's
If this option is ON the changes will affect all the blocks that are in the drawing and all the blocks inserted in the future.

Note, very important: If you only want the changes to affect **future** blocks do not select "Apply changes to existing ref". If you want all blocks changed, select the SYNC button. (see page 9-2)

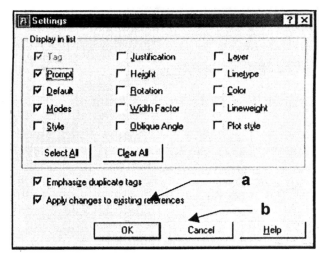

EDIT ATTRIBUTE [Edit...]

The Edit Attribute dialog box has 3 additional tabs, ATTRIBUTE, TEXT OPTIONS and PROPERTIES.

NOTE: If you select the **Auto preview changes** you can view the changes as you make them.

ATTRIBUTE tab
Allows you to change the MODE, TAG, PROMPT and DEFAULT.

TEXT OPTIONS tab
Allows you to make changes to the Text.

PROPERTIES tab
Allows you to make change to the properties.

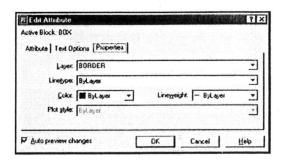

WHAT IF YOU JUST WANT TO EDIT _ONE ATTRIBUTE_ IN _ONE BLOCK?_

The following two commands allow you to edit the attributes in only one Block and it will not affect any other blocks.

Select one of the commands below and then select the block that you wish to change.

To change the Attribute value:

TYPE = ATTEDIT
PULLDOWN = NONE
TOOLBAR = NONE

This command allows you to edit the attribute values for <u>only one block.</u>

This command will not affect other blocks.

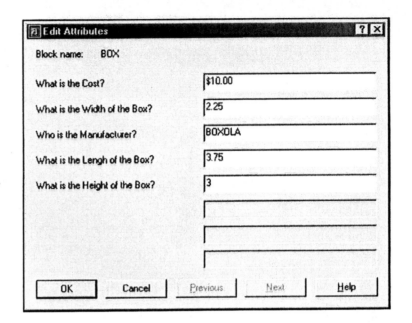

To change the Attribute structure:

TYPE = EATTEDIT
PULLDOWN = MODIFY / OBJECT / ATTRIBUTE / SINGLE
TOOLBAR = MODIFY II

or double click on the block

This command allows you to edit the Attributes, Text Options, Properties and the Values of <u>only one block</u>.

This command will not affect other blocks.

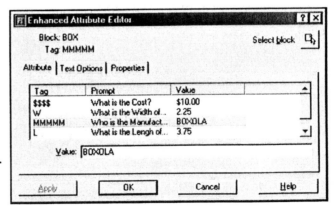

WHAT IF YOU WANT TO <u>EDIT THE OBJECTS IN A BLOCK</u>?

If you would like to add, delete or change objects within an existing block, you may do it easily with the command **Refedit**.
Refedit allows you to make changes to a block, saves the changes and updates all other previously inserted blocks, within the drawing, automatically.
The changes only affect the current drawing.

1. Select the Refedit command using one of the following:

> **TYPE = Refedit**
> **PULLDOWN = MODIFY / XREF & BLOCK EDITING / EDIT REF IN-PLACE**
> **TOOLBAR = MODIFY II**

2. Select the Block you wish to edit. (Click on it, can't use a window)

The Reference Edit dialog box should appear.

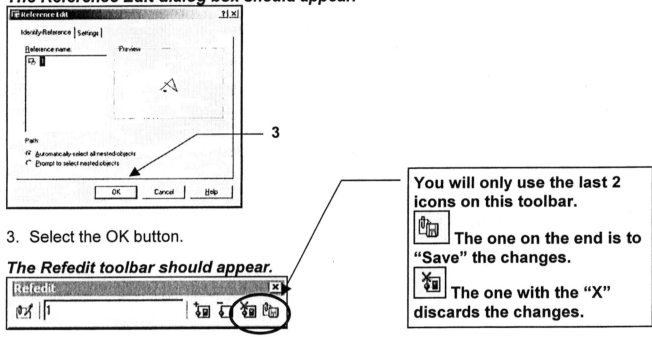

3. Select the OK button.

The Refedit toolbar should appear.

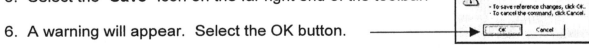

> **You will only use the last 2 icons on this toolbar.**
>
> **The one on the end is to "Save" the changes.**
>
> **The one with the "X" discards the changes.**

Note: The selected block will remain bold but all other objects will fade to gray. This is to emphasis the selected block.

4. Make the changes to the selected block.

Note: As long as the "Refedit" toolbar is open, anything you add, erase or change will affect the selected block.

5. Select the "**Save**" icon on the far right end of the toolbar.

6. A warning will appear. Select the OK button.

The block has been redefined and all of the existing blocks have been updated to reflect the changes you made. (These changes affect the current drawing only)

EXTRACTING ATTRIBUTES

Your blocks can now contain attribute information (data), such as size, manufacturer, cost or maybe even a bill of material. The next step is to learn how to **extract** that **data** and put it into a spreadsheet such as Microsoft Excel.

The first step to extracting attribute data is to create an **Attribute Template** file (.blk). This is a very simple process using the **Attribute Extraction Wizard**.

1. Select the **ATTRIBUTE EXTRACTION WIZARD** using one of the following:

TYPE = EATTEXT
PULLDOWN = TOOLS / ATTRIBUTE EXTRACTION
TOOLBAR = MODIFY II

The Attribute Extraction dialog box should appear.

2. Select one of the following:

   Select Objects: You can select individual blocks within the current drawing.

   Current Drawing: Selects all of the Attributes in the current drawing.

   Select Drawing: You can browse and select multiple drawings.

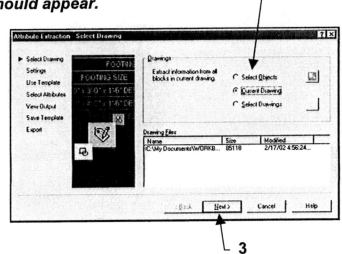

3. Select **Next>**

The Attribute Extraction-Settings dialog box should appear.

You may include blocks with attributes from nested blocks or external reference files. (External Reference files will be discussed in Lesson 11)

If you do not have nested or xref files in the current drawing, it will not hurt to leave these options checked.

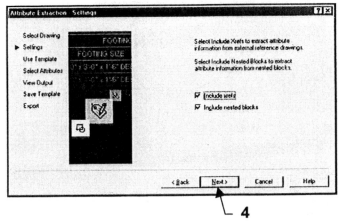

4. Select **Next>**

The Attribute Extraction – Use Template dialog box should appear.

5. Select one of the following:

No Template - When you extract attribute data the information has to be stored somewhere.   So AutoCAD creates a template (.blk) file to store all of the information and instructions you are creating using this Attribute Extraction Wizard.

Use Template - If you have already created an Attribute template file, for the data desired, then select Use template. Then select the template file previously saved.

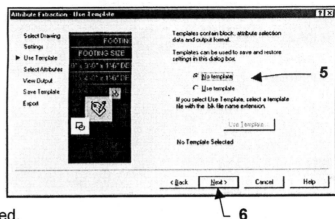

6. Select **Next>**

The Attribute Extraction – Select Attributes dialog box should appear.

With this dialog box you will select exactly what information you wish to extract.
(Relax, this will not be difficult. Just follow the steps below)

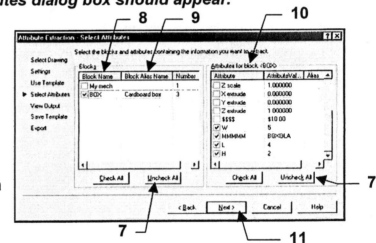

7. **Uncheck All** – To make the selection a lot less confusing, select the Uncheck All buttons <u>under</u> the **Blocks** and **Attributes for Blocks** area. Now you will be able to carefully select exactly what information you want to save.

8. **Blocks** area – Select the block that you would like to extract information from. (In the example above the **Box** block is selected.)

9. **Block Alias Name** – If you would like to change the Block Name, that will be displayed in the data, you can type the name in the **Block Alias Name** box. As an example, you may have a block saved as: **Flr Plan**. You may want **Floor Plan** displayed instead. Click on the block alias cell and type alias name. (In the example above the block name Box has an alias name of Cardboard box.)

10. **Attribute for Block** – Select all the attributes that you would like extracted from the Block that is checked in the Block area. You must do steps 8 and 10 for each block from which you want to extract information. (In the example above, W, MMMM, L and H have been selected)

11. Select **Next>**

The Attribute Extraction – View Output dialog box should appear

The **View Output** dialog box gives you a glimpse of how your final extracted information will look. You may use the Back button to make changes.

Alternate View – You may view this information 2 different ways. The example on the right displays a separate line for each attribute value per block. The other option is a separate listing displaying all attribute values per block on one line.

Copy to Clipboard – Attribute Extraction Wizard is designed primarily to place the information in a data processing program such as Excel. If you would like to put this information in another program, such as Microsoft Word, select the **Copy to Clipboard** button. This will temporarily place the information on the Clipboard so you can paste it into a document.

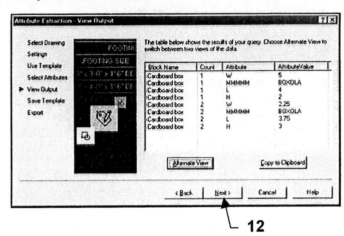

12. Select **Next>**

The Attribute Extraction – Save Template dialog box should appear

Save Template – Save the extracted information as a file.

13. Select the "Save Template" button. Select "where" to save it, give it a name and then select the "Save" button.
The file will be saved with a .blk extension.

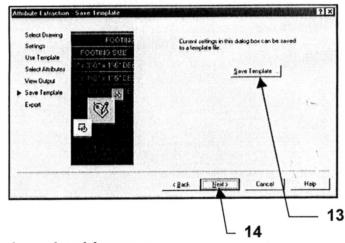

14. Select **Next>**

The Attribute Extraction – Export dialog box should appear

(Now save it again as an Excel file)

15. **File Type** – Select the file type from the drop down list.

16. **File Name** – Type the name or select the browse button to select the location where you want the file saved and name it.

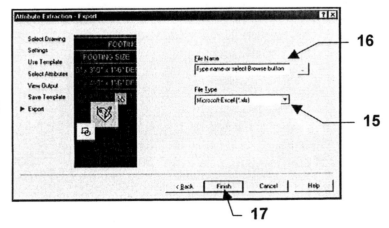

17. Select **Finish**

OPEN THE ATTRIBUTE TEMPLATE FILE IN EXCEL

Now that you have extracted data from your drawing, you will want to insert the data into a data processing program in order to calculate, sort or print out the information.

1. **Start** Excel.

2. Select **File / Open**.

3. **Locate** the Attribute Template file, **select** the file and then **OPEN**.

The file should appear on the screen. Now you can sort, copy, move, print etc.

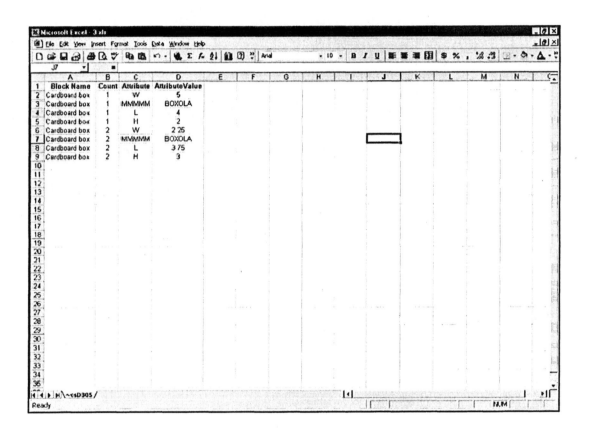

EXERCISE 9A

Extracting Attributes

The following exercise will take you through extracting attributes. You will open an existing drawing and extract the attribute data and then transfer that data to a data base spread sheet.

1. Open **Ex-8B**.
2. Select **Tools / Attribute Extraction**

The Attribute Extractions dialog box should appear.

3. Select **Current drawing**.....then **Next>**

4. Check the **Include Xrefs** and **Nested Blocks** boxes.....then **Next>**

5. Select **No template**......then **Next>**

6. **Blocks** area
   a. Uncheck all
   b. Select **SOFA**

7. **Attributes for Blocks** area
   a. uncheck all
   b. Select **COST, MFR & SIZE**.

8. **Blocks** area
   a. Select **CHAIR**

9. **Attributes for Blocks** area
   a. uncheck all
   b. Select **COST, MFR and SIZE**.

10. **Blocks** area
    a. Select **FILE CAB**
    b. Add Alias: **File Cabinet**

11. **Attributes for Blocks** area
    a. uncheck all
    b. Select **COST, MFR & SIZE**.

12. Select **Next>**

Your Extraction should look approximately like the dialog box shown above.

13. Select **Next>**

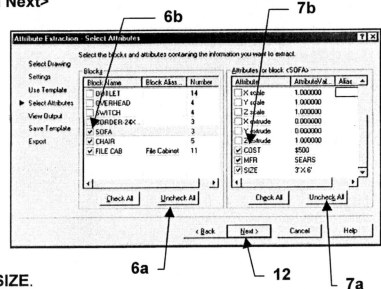

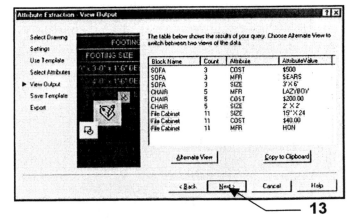

14. Save the template as 9A.
    (This file will have an extension of .blk)

15. Select **Next>**

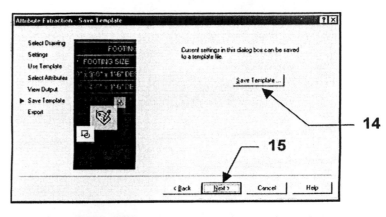

16. Select the **File Type**

    Do not select finish yet!

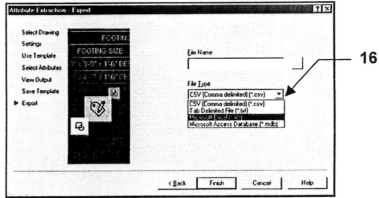

17. Select the **Browse** button to specify where you want to save and what you want to name it.

    Save as 9A again
    (This file will have an extension of .xls)

18. Select **Finish**

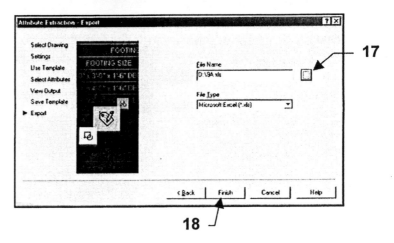

Now you can open the attribute information in an Excel Worksheet.

1. **Start** Excel

2. Select **File / Open**

3. Find the **9A.xls** file and select it.

4. Your worksheet should look approximately as shown.

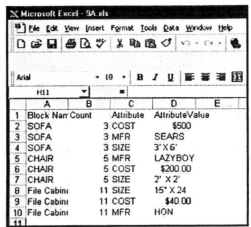

NOTES:

LEARNING OBJECTIVES

After completing this lesson, you will be able to:

1. Navigate in the DesignCenter Palette
2. Open a drawing from the DesignCenter Palette
3. Insert a block from the DesignCenter Palette
4. Drag and drop hatch patterns.
5. Drag and drop Symbols from the Internet
6. Open the Tool Palette Window
7. Control Tool Palette Properties
8, Create a Tool Palette
9. Export and Import a Tool Palette

LESSON 10

DesignCenter

The AutoCAD **DesignCenter** allows you to find, preview and drag and drop Blocks, Dimstyles, Textstyles, Layers, Layouts and more, from the DesignCenter to an open drawing. You can actually get into a previously saved drawing file and copy any of the items listed above into an open drawing.

Opening the DesignCenter palette

To open the **DesignCenter** palette select one of the following:

> **TYPE = DC or Ctrl + 2**
> **PULLDOWN = TOOLS / DesignCenter**
> **TOOLBAR = STANDARD**

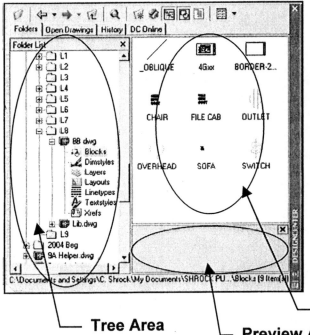

How to:

Resizing – You can change the width and height. Rest the cursor on an edge until the pointer changes to a double ended arrow. Click and drag to desired size.

Dock – Click the title bar, then drag it to either side of the drawing window.

Hide – You can hide the DesignCenter palette using the Auto-hide option. Click on the "properties" button and select "Auto-hide". When Auto-hide is ON, the palette is hidden, only the title bar is visible. The palette reappears when you place the cursor on the title bar.

Content Area

Tree Area

Preview Area

VIEWING TABS

At the top of the palette there are 4 tabs that allow you to change the view. They are: Folders, Open Drawings, History and DC Online.

Folders tab – Displays the Directories and files similar to Windows Explorer. You can navigate and locate content anywhere on your system.

Open Drawings tab – Displays all open drawings. Allows you to select content from an open drawing and insert it into another open drawing. (Note: the target drawing must be the "active" drawing)

History tab – Displays the last 20 file locations accessed with DesignCenter. Allows you to double click on the path to load it into the "Content" area.

DC Online tab – Connects you to the Internet and gives you access to symbols.

BUTTONS

At the top of the DesignCenter palette, there is a row of buttons. (Descriptions below) To select a button, just click once on it.

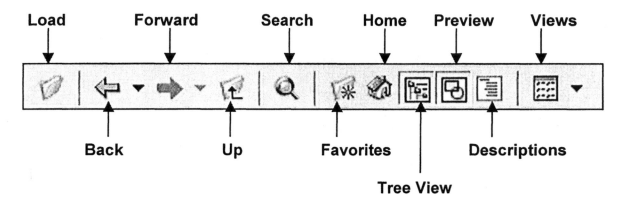

Load – This button displays the "Load" dialog box. (It is identical to the "Select File" dialog box.) Locate the drawing content that you want loaded into the "Content" area. You may also locate the drawing content using the Tree view and Folders tab.

Back and Forward – Allows you to cycle through previously selected file content.

Up – Moves up one folder from the current folder.

Search – Allows you to search for drawings by specifying various criteria.

Favorites – Displays the content of the Favorites folder. Content can be added to this folder. Right click over an item in the Tree View or Content areas then select "Add to Favorites" from the menu that appears.)You may add a drive letter, folder, drawing, layers, blocks etc.)

Home – Takes you to the DesignCenter folder, by default. You can change this. Right click on an item in the Tree view area and select "Set as Home" from the menu. (You many select a drive letter, folder, drawing etc.)

Tree View – Toggles the Tree View On and OFF. Only works when the "Folders" or "Open Drawings" tab is current.

Preview – Toggles the "Preview" area On and OFF.

Description – Toggles the "Description" area On and OFF. (A description must have been given at the time the block was created.)

Views – Controls how the "Content" area is displayed. The choices are: Large or small icons, a list view or a detailed view.

How to open a drawing from DesignCenter Palette.

1. Open a "New" drawing, "Start from Scratch".
   (Typically you would want to "open" a drawing into a "new" drawing. But you may also open a drawing into an already open drawing)

2. Locate the drawing you want to open.

3. Make sure it is in the "Content" area. (You cannot open a drawing from the Tree area.)

4. Click and drag the drawing onto the blank drawing area.

5. You will be prompted for the insertion point. Typically you should use 0, 0 .

6. You will then be prompted for the scale for X and Y and then the rotation angle.

Does this sound familiar. Yes, this is just like inserting a block. In fact, the drawing has actually been inserted as a block. If you want to modify the drawing you must Explode it.

How to Insert a Block using DesignCenter

There are two methods of inserting a block using DesignCenter.

Method 1: Drag and Drop
Locate the Block in the Content area and drag and drop it into the drawing area of an open drawing. You will not be prompted for the insertion point.

Method 2: Specified Coordinates, Scale and Rotation
Double click on the block in the Contents area. The Insert dialog box will appear. Specify coordinates for insertion point, scale factor and rotation angle.

To Close the DesignCenter palette

The DesignCenter palette will remain open until you close it.

(If you have "Autohide" ON, only the title bar will remain visible)

To Close the DesignCenter palette, select the "X" in the upper right corner of the Palette.

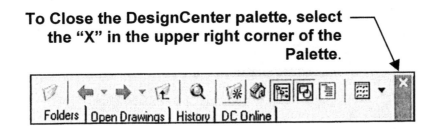

Drag and drop Hatch Patterns

AutoCAD allows you to drag and drop hatch patterns from the DesignCenter directly into a closed area in your drawing.

1. Open the DesignCenter.

2. <u>Find and select</u> AutoCAD's "**acad.pat**" file located in the **Support** directory.
    (You may have to do a Search to locate it on your system.)
    AutoCAD's hatch patterns are stored in this file.

3. You should be able to see the hatch pattern squares in the <u>Content area</u>.

4. **Important**: First select the layer that you want the hatch pattern to reside on.

5. Now simply click on a hatch pattern square and drag it on to your drawing and drop it on the area where you want the hatch.

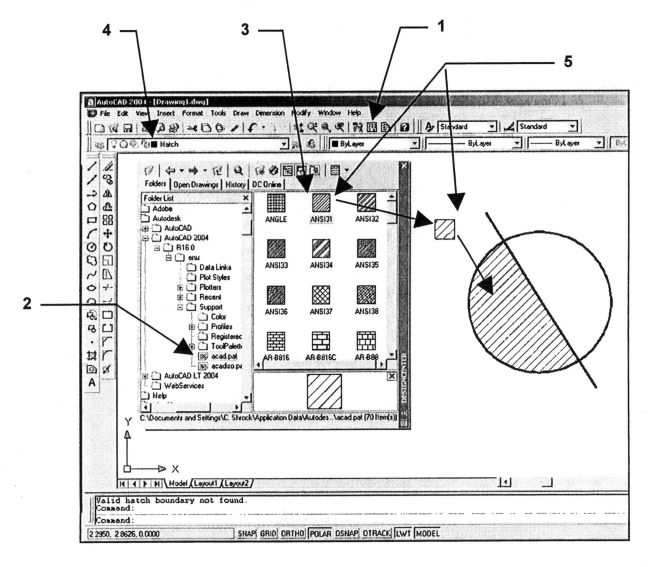

Drag and drop Layouts, Layers, Text Styles etc.

You can drag and drop just about anything from an existing file, listed in the Tree view "Folder List", to an open drawing file. This is a significant time saver. Just locate the source file in the tree area and drag and drop into an open drawing.

Some examples are shown below.

BLOCKS

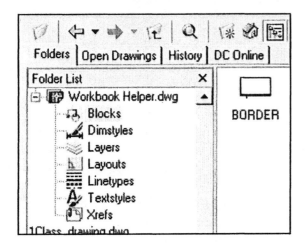

DIMENSION STYLES

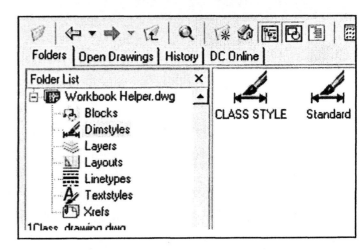

LAYERS

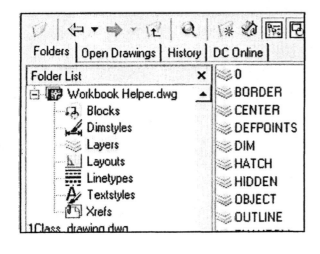

LAYOUTS

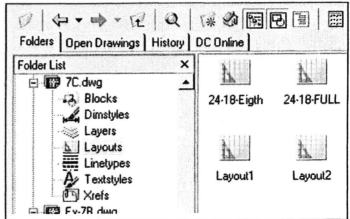

DC Online

DC Online connects you to the Internet and gives you access to thousands of symbols and manufacturer's product information.

Note: You must have an Internet connection to use this feature.

Just select the "DC Online" tab and AutoCAD automatically opens your Internet connection to the correct Internet address.

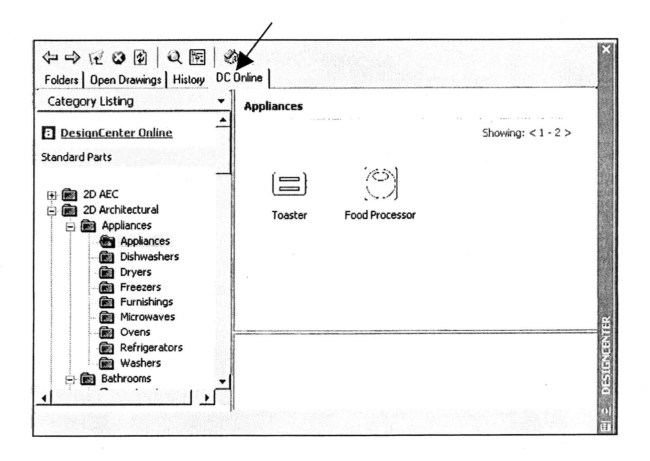

TOOL PALETTES

AutoCAD has developed another tool to use for storing and inserting frequently used features.   This tool is a "**Tool Palette**".

How to Open the Tool Palette

To open the Tool Palette, select one of the following:

> **TYPE = toolpalettes or Ctrl + 3**
> **PULLDOWN MENU = Tools / Tool Palettes Window**
> **TOOLBAR = STANDARD**

The following palette should appear.

The entire palette is called the **Tool Palettes Window.**

The individual tabs are called **Tool Palettes or Palettes.**
(AutoCAD comes with 3 default tool palettes that you can delete completely or add additional tools.)

Any icon on a tool palette is called a **Tool.**

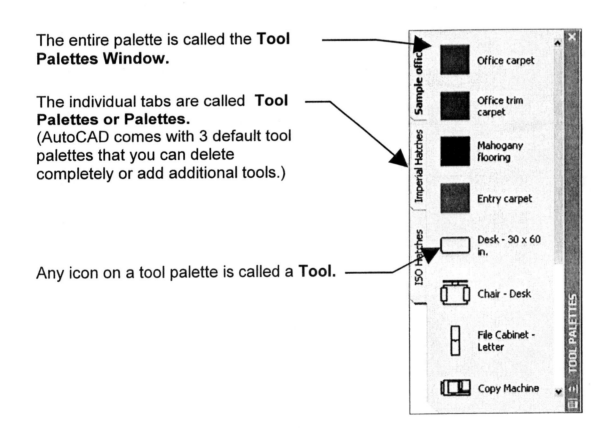

PALETTE PROPERTIES

You can control the appearance and operation of a Tool Palette.

Right click inside the palette. The menu shown below should appear:

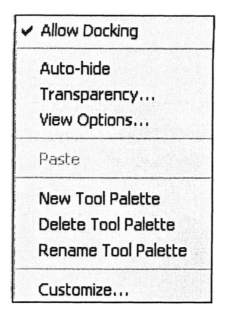

ALLOW DOCKING
This feature allows you to dock the Tool Palettes window on the left or right side of the AutoCAD drawing area. Note: A docking window will never cover your drawing, but it will reduce the drawing area.

AUTO-HIDE
When Auto-hide is ON (with check mark), the Tool Palettes Window is hidden and only the title bar is visible. The Tool Palettes Window reappears when you place the cursor on the title bar. When Auto-hide is OFF (no check mark), the Tool Palette Window remains fully visible.

TRANSPARENCY
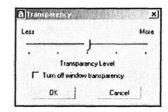

Use this option to make the Tool Palette window transparent so you can "see through" the window to the drawing underneath. Slide the pointer to the desired level of transparency.

Note:
1. *Transparency feature is available only when hardware acceleration is off. To select Software acceleration, select Tools / Options / System tab. Select the Properties button from the Current 3D Graphics Display section. In the Acceleration section, select Software. Apply and Close.*
2. *Transparency is only available when the palette is not docked.*
3. *Transparency is not available if you are using the Micorsoft Windows NT operating system.*

Note: Transparency may slow performance. You may prefer to turn it off.

VIEW OPTIONS

This option allows you to specify the *"Image Size"* and "View Style". Also you need to specify whether to apply the options to the "Current Tool Palette" or "All Tool Palettes".

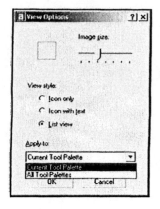

Icon only – Icons with no text

Icon with Text – icons with text in multiple columns.

List View – icons with text always in a vertical column.

NEW TOOL PALETTE

This option allows you to create a new Tool Palette. This option is explained on page 10-11.

DELETE TOOL PALETTE

This option allows you to delete an existing Tool Palette. First select the Tool Palette to be deleted, right click and select "Delete Tool Palette". A warning will be displayed basically saying "Are you sure?". If you select the OK button, the Tool Palette will be permanently deleted.

RENAME TOOL PALETTE

This option allows you to rename an existing Tool Palette. First select the Tool Palette to be renamed, right click and select "Rename Tool Palette". Enter the new name in the box displayed and <enter>. The new name will be shown on the tab.

CUSTOMIZE

This option allows you to move the Tool Palette "tabs" up or down. This changes the order in which they are displayed. You might use this to move a more frequently used Tool Palette to the front.

*The **Import** and **Export** features, shown on this dialog box, are discussed on page 10-12.*

HOW TO MOVE OR COPY A TOOL TO ANOTHER PALETTE
1. Click the palette tab that contains the tool you want to move or copy. (Source)
2. Right click the tool you want to move or copy.
3. Choose "Cut" (to move) or "Copy" (to copy)
4. Click the palette tab on which you want to place the item. (Target)
5. Right click any blank area on the palette and choose "Paste".

HOW TO CREATE A TOOL PALETTE

Creating a new tool palette means <u>adding a "tab" to the Tool Palettes Window</u>.

3 types of palettes can be created:
> **Drawing** – Inserts drawings as blocks
> **Blocks** – Inserts blocks
> **Hatch** - Inserts hatch patterns

How to creating an "<u>Empty</u>" tool palette
1. Right click in the Tool Palettes window
2. Choose "New Tool Palette" from the menu.
3. Enter a name in the box that appears

How to create a tool palette and its content using the DesignCenter
1. Open the DesignCenter
2. Locate one of the following:
   a. Folder – to add all the drawings in the folder
   b. Drawing file – to add all the blocks in the drawing
   c. Block icon – to add the block
   d. Hatch icon - to add all hatches in the .pat file
3. Right click the item and choose "<u>Create Tool Palette</u>"

(If you select a folder, choose "Create Tool Palette of Blocks". If you select a hatch file, choose "Create Tool Palette of Hatch Patterns")

Tool Properties
Before you start using your new tools, you should set their properties. Tool properties specify how a drawing, block or hatch is inserted. For example, you can specify that a hatch be inserted at a certain scale or a block be inserted on a specific layer.

To set Tool properties, right click any tool and choose Properties from the menu. The Tool Properties dialog box will varies slightly depending on whether it is for a block, drawing or hatch pattern.

Updating the Icon for a Tool
If you change a Block you must update its icon in the Tool Palette. The easiest way to accomplish this is "delete" the icon and replace it using the DesignCenter. The more difficult way is to Right Click on the tool and delete the "Source File" field and enter the correct path.

Block Master Drawing
I like to create and maintain a "Master" drawing for all of my blocks. I named it "Library". I create all of my blocks in the drawing. If I need to edit the block, I do it within the "Library" drawing. If I want to add a block to a Tool Palette, I know exactly where to find the block using DesignCenter. The "path" to the original Tool never changes unless I change the location of my "Library" drawing.

HOW TO EXPORT TOOL PALETTES

1. Right click on the Tool Palette.
2. Select "Customize" from the menu.
3. Select the "Export" button.
4. Select the where to save the Tool Palette. (Save in)
5. Enter a name for the Tool Palette.
6. Select the " Save " button. (The Tool Palette is saved with an .xtp extension)

HOW TO IMPORT TOOL PALETTES

1. Right click on the Tool Palette.
2. Select "Customize" from the menu.
3. Select the "Import" button.
4. Locate the Tool Palette to import.
5. Select the "Open" button.

Note: Tool Palettes are not saved with the drawing file. They are saved as part of the AutoCAD profile. So exporting and importing are very useful tools. If you always use the same computer your specific set of Tool Palettes will always appear. But if you move to another computer it will be necessary for you to Import your Tool Palettes to your Profile. (For Profiles, refer to page 2-8)

Summary
Using "Create a new Toolbars" (page 2-3) for drawing commands and "Create new Tool Palettes" (page 10-11) for blocks, then adding both to your "Profile" (page 2-8) enables you to customize your AutoCAD work area for maximum accessibility.

EXERCISE 10A

Inserting Blocks from the DesignCenter.

1. Open **EX-5E or 5F** (depending on which one you completed).

2. Make blocks of the chair, desk, sofa, stand with lamp, door, window and file cabinet in the boss's office.

   These blocks must be saved __in__ the current drawing file. Use __Retain__ or __Delete__ when selecting objects. Do not use __Convert to block__.

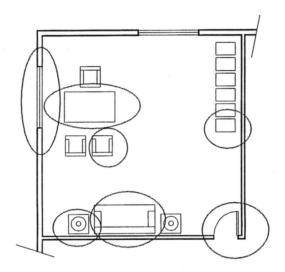

3. **Save** the drawing as **EX-10X (Note: 10X will not appear in the tree until you close and re-open the directory in the DesignCenter)**
4. **Open** My Feet-Inches Setup and select the **Qtr Equals Foot** tab.
5. Draw the simple floor plan shown below. (Walls = 6" wide)
6. **Save** as **EX-10A** and continue on to page 10-14.

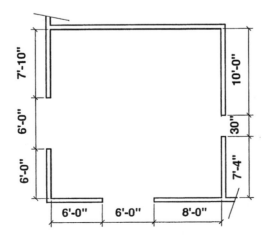

7. **Open** the **DesignCenter** (Refer to page 10-2 if necessary)
8. **Find** your drawing **EX-10X** in Tree View (left side)
9. Select the **plus box** beside the file to display the contents of EX-10X.
10. Select **"Blocks"** to display the blocks in the Content area. (right side)

Note: Experiment with the 4 ways you can display the blocks. Large icon, small icon, List, and details. (Refer to page 10-3, Views)

12. **Insert** the Blocks from the Content area to create the drawing below.
    (Refer to page 10-4, How to Insert a Block using DesignCenter, if necessary)
    a. Use the **drag and drop** method for the blocks that are not rotated.
    b. Use **Specify Coordinate** method for the blocks that are rotated.

13. Save as EX-10A.

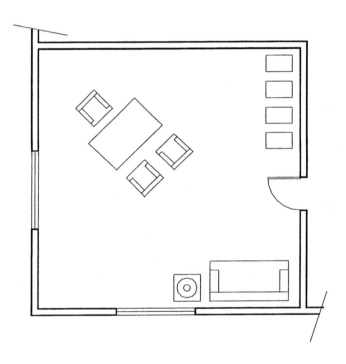

EXERCISE 10B

Inserting Symbols from the DC Online.

1. Open **EX-10A.**
2. Select the **Furniture** layer.
3. Open the DesignCenter.
4. Select the **DC Online** tab.
5. Locate a telephone plan view in the following directory and select it:
   2D Architectural / Telephone / Modems / Modem Plan View
6. Drag and drop the symbol into the open drawing.
7. You will be prompted for an insertion point, first enter "R" for **Rotate.**
8. Enter the Angle: **45 <enter>.**
9. Now place the insertion point on the desk.

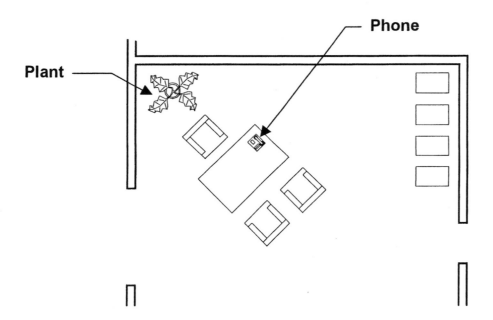

10. Select the **Misc** layer.
11. Locate a Tree Palm in the following directory and select it:
    2D Architectural / Landscaping / Trees / Tree Palm
12. Drag and drop the symbol into the open drawing.
13. You will be prompted for an insertion point, first enter "S" for **Scale.**
14. Enter the Scale: **.3 <enter>.**
15. Now place the insertion point in the corner.
16. Save as **EX-10B**

EXERCISE 10C

Inserting a Hatch pattern from the DesignCenter.

1. Open **My Decimal Setup**

2. Draw the objects shown below.

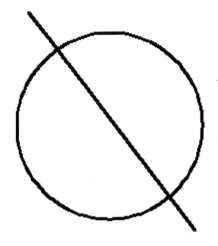

3. Open the DesignCenter

4. Locate AutoCAD's **acad.pat** files (Refer to page 10-5)

5. Apply Ansi 31 pattern on the right side and Ansi 32 on the left side. Set the rotation angle of Ansi 32 to appear as the example below.

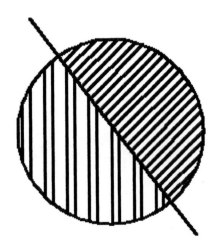

6. Save as **Ex-10C**.

EXERCISE 10D

Create a Tool Palette

1. Open **My Decimal Setup**

2. Create a new empty Tool Palette.
   a. Name = **My Palette**

3. Add a Hatch pattern Ansi 31

4. Add a block from DesignCenter / Kitchens.dwg / Dishwasher

5. Move the "My Palette" Tool Palette tab name up to the top.

Your Tool Palette window should look like the Tool Palette Window shown below.

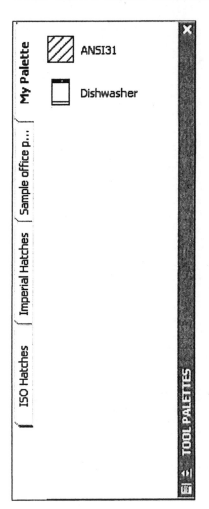

6. Save as **Ex-10D**.

7. Export this Palette to a disk.

NOTES:

LEARNING OBJECTIVES

After completing this lesson, you will be able to:

1. Understand the use of Externally Referenced drawings
2. Insert an Xref drawing
3. Use the Xref Manager
4. Bind an Xref to a drawing.
5. Clip an Xref drawing.
6. Edit an Xref drawing using the XOPEN command.
7. Convert an object to a Viewport
8. Create multiple viewports and multiple xrefs
9. Understand the PSLTCALE command.
10. Create multiple viewports quick and easy.

LESSON 11

External Reference drawings (xref)

The **XREF** command is used to insert an image of another drawing into the current drawing. This command is very similar to the INSERT command. But when you externally reference (XREF) drawing "A" into drawing "B", the image of drawing "A" appears but only the **path** to where the original drawing "A" is stored is loaded into drawing "B". Drawing "A" does not become a permanent part of drawing "B".

Each time you open drawing "B", it will look for drawing "A" (via the path) and will load the image of the **current version** of drawing "A". If drawing "A" has been changed, the new changed version will be displayed. So drawing "B" will always display the most current version of drawing "A".

Since only the path information of drawing "A" is stored in drawing "B", the amount of data in drawing "B" does not increase. This is a great advantage.

If you want drawing "A" to actually become part of drawing "B", you can **BIND** "A" to "B". (Refer to page 11-6)

If you want the image of drawing "A" to disappear temporarily, you can **UNLOAD** it. To make it reappear simply **RELOAD** it. (Refer to page 11-5 & 6)

If you want to delete the image and the path to drawing "A", you can **DETACH** it. (Refer to page 11-5)

Examples of how the XREF command could be useful?

1. **If you need to draw an elevation**. You could XREF a floor plan and use it basically as a template for the wall, window and door locations. You can get all of the dimensions directly from the floor plan by snapping to the objects. It will not be necessary to refer back and forth between drawings for measurements. If there are any changes to the floor plan at a later date, when you re-open the elevation drawing, the latest floor plan design will appear automatically. To make the floor plan drawing become invisible, simply UNLOAD it. When you need it again, RELOAD it.

2. **If you want to plot multiple drawings on one sheet of paper.** Many projects require "standards" or "detail" drawings that are included in every set of drawings you produce. Currently you probably drew each detail on the original drawing package, or created those transparent "sticky backs". Now, using the XREF command, you would do the following. 1. Make individual drawings of each standard or detail. 2. XREF the standard or detail drawing into the original drawing package. This would not only decrease the size of your file but if you made a change to the standard or detail, the latest revision would automatically be loaded each time you opened the drawing package.

3. **Working with a team of drafters all working on a segment of the project.** (This process works best if your office is networked.) Let's say you are responsible for the furniture layout on an architectural project and the floor plan is not quite finished. You could XREF the current version of the floor plan and start working on the furniture layout. Every couple of hours you can RELOAD the XREF floor plan and your drawing will be automatically updated to the latest work saved on the floor plan.

HOW TO INSERT AN XREF DRAWING

1. Open the drawing into which you will be externally referencing another drawing. We will call it drawing "B". (For example, open My Decimal Setup)

2. Select a Layout tab.

3. Select a Viewport. (double click inside the viewport)
   Note: The viewport can be locked or unlocked, it does not matter.

4. Select the XREF command using one of the following:

TYPE = XATTACH
PULLDOWN = INSERT / EXTERNAL REFERENCE
TOOLBAR = REFERENCE

The following "Select Reference File" dialog box should appear.

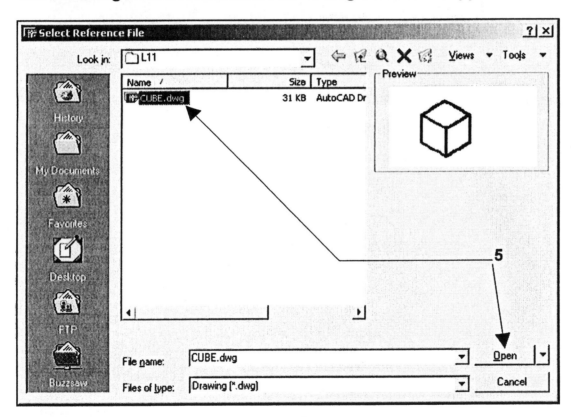

5. Select the file you wish to **XREF** and select **Open.**

continue on the next page….

*The following "**External Reference**" dialog box should appear.*

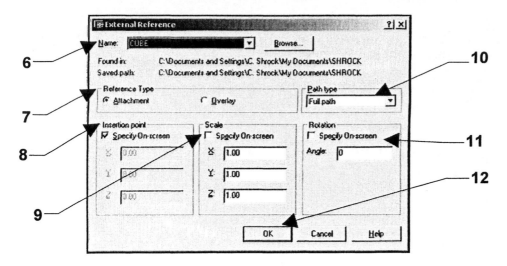

6. The name of the file you selected should appear in the Name box.

7. Select **"Reference Type".**

    Attachment - This option attaches one drawing to another. If you attach drawing A to drawing B, then attach drawing B to drawing C, drawing **A remains attached**. (Daisy Chain)

    Overlay - This option is exactly like Attachment except, if you Overlay drawing A to drawing B, then attach drawing B to drawing C. Only drawing B remains attached, drawing A does not.

8. **Insertion Point -** Specify the X, Y and Z insertion coordinates in advance or check the box "Specify on-Screen" to locate the insertion point with the cursor.

9. **Scale -** Specify the X, Y and Z scale factors in advance or check the box "Specify on-screen" to type the scale factors on the command line.

10. **Path -** Select the Path type to save with the xref.

11. **Rotation -** Specify the Angle or check the box "Specify on-screen" to type the scale factors on the command line.

12. Select the **OK** button.

13. Place the insertion point with your cursor if you have not preset the insertion point.

Note: If you xref a drawing and it does not appear, try the following:
1. Verify that you xreffed the drawing into Model space, not paper space.
2. Unlock the viewport, use View / Zoom / Extents to find the drawing in a viewport, adjust the scale and re-lock the viewport.
3. Go back to the original drawing to verify that it was drawn in model space. Objects drawn in paper space will not xref.
4. Go back to the original drawing and verify that the layers are not frozen.

XREF MANAGER

The **XREF MANAGER** allows you to view and manage the xref drawing information in the current drawing.

Select the XREF Manager using one of the following:

TYPE = XREF
PULLDOWN = INSERT / XREF MANAGER
TOOLBAR =

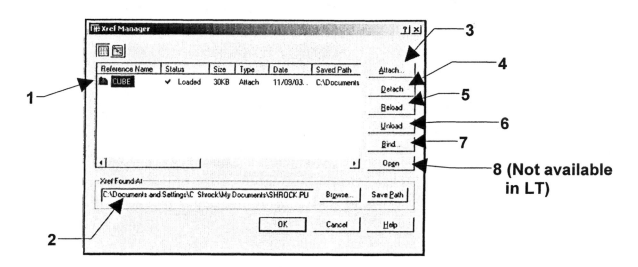

1. **Title Area** - This describes each individual xref drawing in the current drawing.
   Reference Name = The original drawing file name of the xref drawing
   Status = This lists whether the drawing was loaded, not found or unloaded.
   Size = Indicates the size of the xref drawing.
   Type = Lists whether the xref drawing was referenced as an Attachment or Overlay.
   Date = Lists when the xref drawing was originally referenced.
   Saved Path = This is the path the computer will follow to find the xref drawing listed and load it each time you open the current drawing.

2. **Xref Found At:** - When you highlight an xref drawing in the list, the path is shown here. This gives you the opportunity to change the path here if the original xref drawing has been moved and the path is no longer correct.

3. **Attach** - Takes you back to the External Reference dialog box so you can select another drawing to xref.

4. **Detach** - This will remove all information about the selected xref drawing from the current file. The xref drawing will disappear immediately. Not the same as Unload.

5. **Reload** - This option will reload an unloaded xref drawing or update it.
   If another team member is working on the drawing and you would like the latest version, select the Reload option and the latest version will load into the current drawing.

6. **Unload** - <u>Unload is not the same as Detach</u>. An unloaded xref drawing is not visible but the information about it remains and it can be reloaded at any time.

7. **Bind** – When you bind an xref drawing, it becomes a permanent part of the current base drawing. All information in the Xref manager concerning the xref drawing selected will disappear. When you open the base drawing it will not search for the latest version of the previously inserted xref drawing.

8. **Open** - Opens the selected xref for editing in a new window. The new window is displayed after the Xref Manager is closed.

When a drawing is XREFed, its Blocks, Layers, Linetypes, Dimension Styles and Text Styles are kept separate from the current drawing. The name of the xref drawing and a pipe (I) symbol is automatically inserted as a prefix to the newly inserted xref layer. This assures that there will be no duplicate layer names.

Layers belong to host dwg.

Layers belong to Xref dwg.

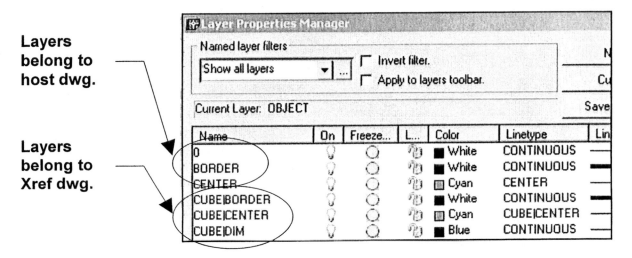

XBIND

It is important to understand that these new names are listed but they cannot be used. Not unless you bind the entire xref drawing or use the Xbind command to bind individual objects.

The following is an example of the Xbind command with an individual layer. You may also use this command for Blocks, Linetypes, Dimension Styles and Text Styles.

1. Type **XBIND** <enter> at the command line.

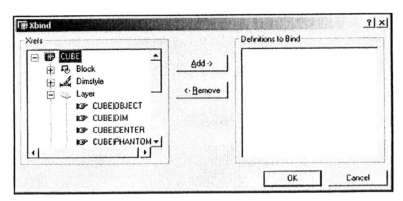

2. Expand the Xref file information list by clicking on the **+** sign beside the Xref drawing file name; then click on the **+** sign beside "Layer". The + sign will then change to a - sign, as shown below.

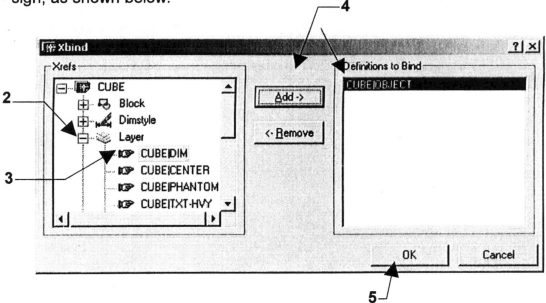

3. Select the layer name to bind.

4. Select the **ADD** button. The xref layer will now appear in the "Definitions to Bind" window.

5. Select the **OK** button.

6. Now select Format / Layers and look at the xref layer. The Pipe (I) symbol between the xref name and the layer name has now changed to 0.
   This layer is now usable.

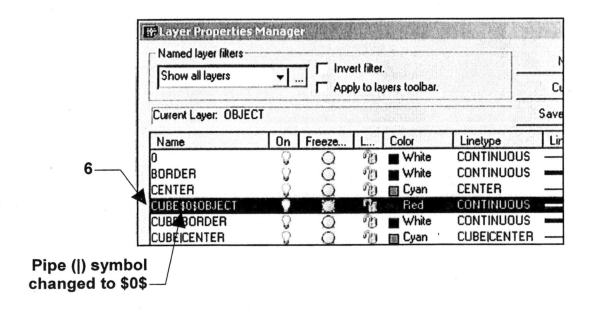

Pipe (|) symbol
changed to 0

Clipping an External Reference

When you Xref a drawing, sometimes you do not want the entire drawing visible. You may only want a portion of the drawing visible. For example, you may Xref an entire floor plan but actually only want the kitchen area visible within the viewport.

This can be easily accomplished with the Xclip command. After you have inserted an External Referenced drawing, select the Xclip command. You will be prompted to specify the area to clip by placing a window around the area. All objects outside of the window will disappear.

How to use the **Xclip** command

1. Select the **Xclip** command using one of the following:

   TYPE = XC
   PULLDOWN MENU = MODIFY / CLIP / XREF
   TOOLBAR = MODIFY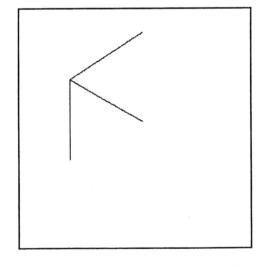

2. Select the Xref drawing to clip.

   Enter clipping option
   [ON/OFF/Clipdepth/Delete/generate Polyline/New boundary] <New>: *N <enter>*

   Specify clipping boundary:
   [Select polyline/Polygonal/Rectangular] <Rectangular>: *R <enter>*

   Specify first corner: *select the location for the first corner of the "Rectangular" clipping boundary.*

   Specify opposite corner: *select the location for the opposite corner of the "Rectangular" clipping boundary.*

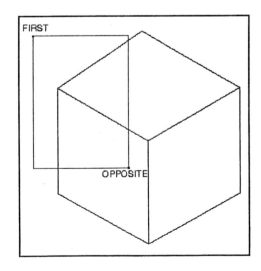
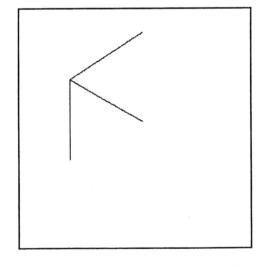

CLIPPING OPTIONS

After you select the Xclip command, the following prompt, with options, appears.

**Enter clipping option
[ON/OFF/Clipdepth/Delete/generate Polyline/New boundary] <New>:**

The description for the options are shown below:

> **ON / OFF** – If you have "clipped" a drawing, you can make the clipped area visible again using the "OFF" option. To make the clipped area invisible again, select the "ON" option.

> **CLIPDEPTH** – This option allows you to select a front and back clipping plane to be defined on a 3D model.

> **DELETE** – To remove a clipping boundary completely.

> **GENERATE POLYLINE** – This option creates a polyline object to represent the clipping border of the selected Xref. (The boundary is invisible by default)

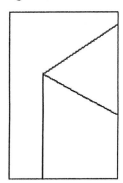

> **NEW BOUNDARY** – Allows you to select a new boundary.

After the New Boundary option is selected, the following prompt appears.

**Specify clipping boundary:
[Select polyline/Polygonal/Rectangular] <Rectangular>:**

The description for the options are shown below:

> **SELECT POLYLINE** – Allows you to select an existing polyline as a boundary.

> **POLYGONAL** –Allows you to draw and irregular polygon, with unlimited corners, as a boundary.

> **RECTANGULAR** – Allows a rectangular shape, with only 2 corners, for the boundary.

EDIT AN EXTERNAL REFERENCED DRAWING

An Xref drawing can be edited very easily using the Xopen command. The Xopen command allows you to open the Xref drawing in a separate window, make the changes and save those changes to the original of the Xref drawing. Just "Reload" to view the changes in your drawing.

(Note: This command is not available in AutoCAD LT)

How to use the Xopen command.

1. Select the **Xopen** command using:

    TYPE = XOPEN

2. Select the Xref drawing, within the host drawing, to be changed.

    The Xref original drawing will open in a separate window.
    (The "Single-drawing compatibility mode" must be "off", unchecked.
    Ref. page Intro-7)

3. Make the necessary changes.

4. Save the drawing. *(Note, you must use the same name.)*

5. Close the drawing, using **File / Close**.

    Your host drawing will reappear with a "balloon message" notifying you that
    the Xref drawing has been changed and it needs to be "Reloaded".

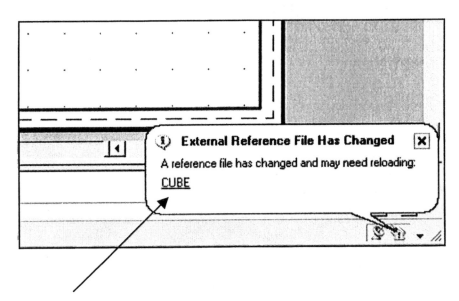

6. Click on the drawing name (underlined and blue).

The following dialog box should appear.

7. Select the drawing to reload. *(Notice the blue check mark and the red exclamation mark. This indicates which drawing needs reloading.)*

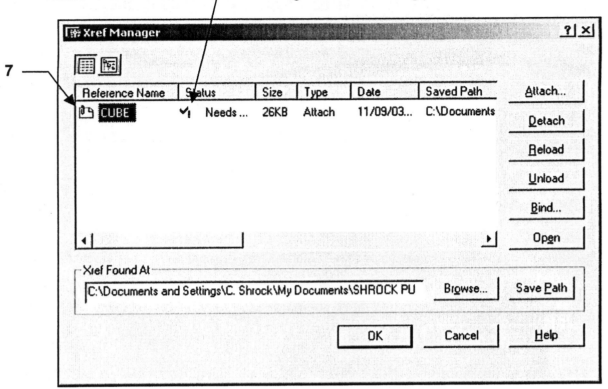

8. Select the **Reload** button. *(Notice the check and exclamation mark changed to rotationing green arrows. This indicates that the drawing has been reloaded)*

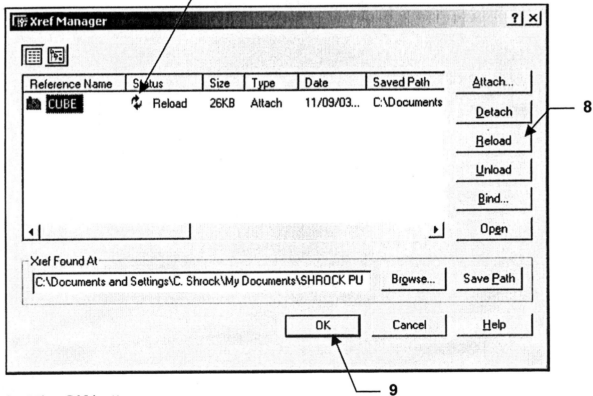

9. Select the **OK** button.

Convert an Object to a Viewport

You may wish to have a viewport with a shape other than rectangular. AutoCAD allows you to convert Circles, Rectangles, Polygons, Ellipse and Closed Polylines to a Viewport.

(Note: This option is not available in AutoCAD LT)

1. You must be in Paperspace.
2. Draw the shape of the viewport using one of the objects listed above.

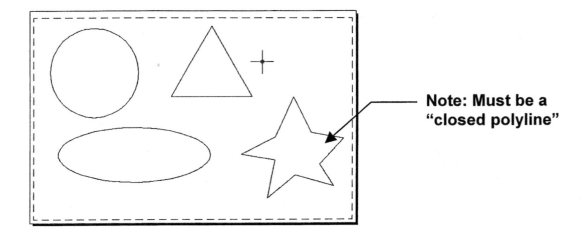

Note: Must be a "closed polyline"

3. Select the Convert Object to Viewport icon on the Viewports toolbar **or** View / Viewports / Object.

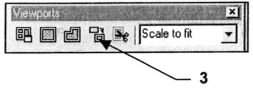

3

4. Select the object to convert.

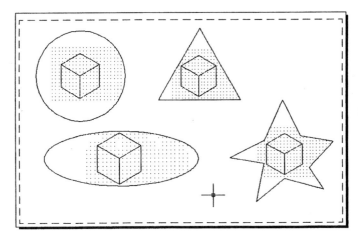

Now you can see through to model space. This is like looking out all the windows in your living room and seeing the same tree growing in your front yard.

Creating Multiple Viewport and Multiple Xrefs

When using Xrefs you will usually want multiple drawings Xrefed. And you will also want multiple viewports. It will not be difficult to understand how to create multiple viewports but you will have to think a little about multiple xrefs.

The following is a example of how to control the viewing of multiple xrefs. The actual exercise, with step by step instructions, is EX-11A

1. Open My Decimal Setup

2. Select the 11 X 17 (1 to 1) tab.

3. Delete any viewports that already exist. (Click on the frame then Erase)

4. Select the viewport layer.

5. Create 2 viewports approximately as shown below.
   First one, than the other.
   Simple so far, huh?

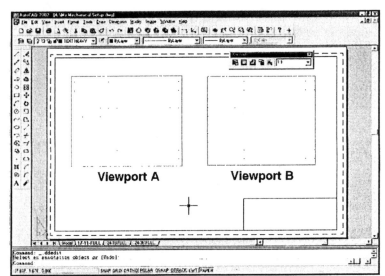

6. Double click inside the Viewport A, on the left.

7. Select the XREF layer.

8. Xref a drawing into this viewport. (7D would be a good one to use)

9. Select Zoom / Extents (You should see the entire drawing inside Viewport A)

10. Now activate Viewport B (double click inside Viewport B)

11. Select Zoom / Extents (You should see the entire Xref drawing inside Viewport B also)

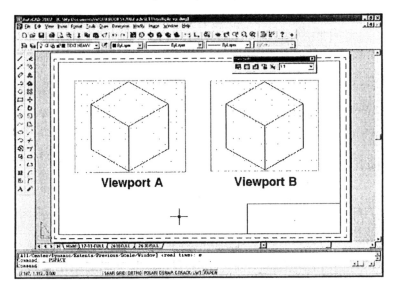

Now take a minute to think about this. This is an important concept to understand. *Pretend that the Viewport frames are 2 windows in your living room and the xref drawing is a tree in your front yard. If you stand in front of one of the windows you can see the tree. Then if you walk to the other window you see the same tree.*

This is basically what is happening with AutoCAD. You have xrefed a drawing into Model space (your front yard) and you are in paperspace (your living room) looking through the viewports (your windows) to model space.

Now let's make the xref in Viewport B disappear.

If you do not want to see the xref drawing in Viewport B you <u>must "**Freeze in Current Viewport**" the layers of the xrefed drawing inside Viewport B.</u> *This is easier than is sounds.*

12. Activate Viewport B. (click in it)

13. Select Format / Layer

Scroll up and down and look at the layer names. Notice some layers have been added. (Refer to page 11-6)

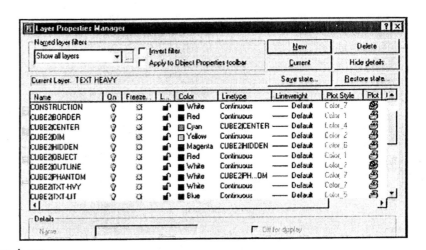

When the drawing was Xrefed, the layers came with that drawing.

You can view the specific layers that belong to the Xref drawing by selecting the xref drawing name in the **Named Layer Filters** box.

14. Select all of the layers that belong to the XREF drawing. (highlight them)

15. Select the "Freeze in Current Viewport" box.

16. Select OK

17. The Xrefed drawing should have disappeared in Viewport B. If it didn't do items 12 through 16 again.

18. Now you should adjust the scale in Viewport A and lock it.

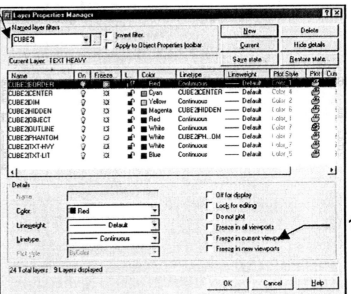

15

Now think about this. If you were to Xref another different drawing into Viewport B, it would also be visible in Viewport A (just another tree in your front yard). So you would have to activate Viewport A, select the layers that belong to the newly Xrefed drawing and select "Freeze layers in Current Viewport". Then only one drawing would be visible in each viewport.

CREATING MULTIPLE VIEWPORTS – A QUICK METHOD

1. Select **VIEW / VIEWPORTS / NEW VIEWPORTS**

2. Select **Four: Equal** from the column on the left.
   This will divide the paperspace into 4 equal viewports

3. Enter **.50 Viewport Spacing**
   This the spacing between each of the new viewports.

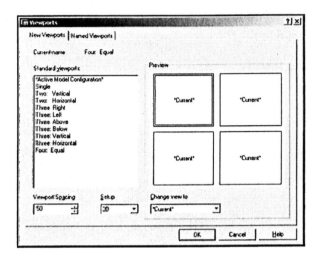

4. Select the OK button.

5. The following prompt will appear:

   Specify first corner or [Fit] <Fit>: ***Enter the location of the first corner of the area to divide and then the opposite corner.***

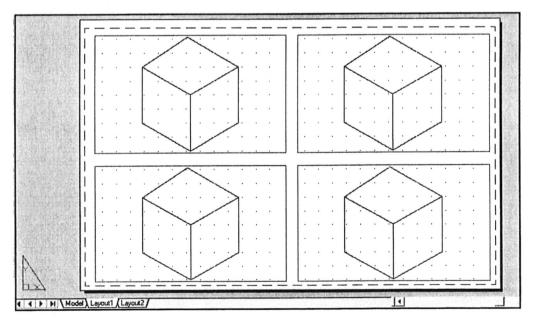

Note: If you have objects in Modelspace, they will appear in each viewport.

11-15

Non-continuous Linetype scales within Viewports

When you adjust the scale within a viewport, the image inside the viewport becomes larger or smaller. This you already know. But what you probably didn't notice is the non-continuous linetypes also change in scale. AutoCAD has a command that will control the linetype scale, within each viewport and paper space, based on the viewport scale or the paper space scale. This command is PSLTSCALE. (Paper space linetype scale) You may control the appearance of these linetypes by setting the PSLTSCALE to 1 or 0. (on or off)

PSLTSCALE set to 0 (off)
The linetype scale is based on the scale of the individual viewports and paperspace. All non-continuous linetype (Center, dashed, etc.) can appear different if the scales in the individual viewports and paper space are different.

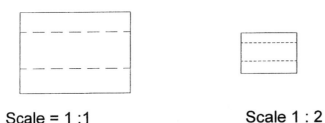

Scale = 1 :1   Scale 1 : 2

PSLTSCALE set to 1 (on; default setting)
The linetype scales for both model space and paper space are scaled to paper space units. **All non-continuous linetypes will appear equal**.

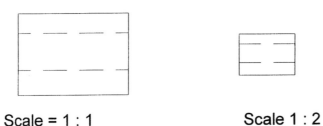

Scale = 1 : 1   Scale 1 : 2

Note: After you change the PSLTSCALE you must select View / REGENALL to display the effects of the new setting.

To change PSLTSCALE:

Command: PSLTSCALE <enter>
Enter new value for PSLTSCALE: *type 1 or 0 <enter>*
Command: REGENALL <enter>

EXERCISE 11A

XREF MULTIPLE DRAWINGS

1. Open **My Decimal Set Up**

2. **Draw** a Rectangle (5" L X 3" W) <u>in models space</u> and **save** as **"RECT"**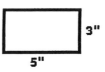

3. **Close** the drawing.

4. **Open** My Decimal Set Up **again**.

5. **Draw** a Circle (2" Radius) <u>in model space</u> and **save** as **"CIRCLE"**

6. **Close** the drawing.

7. Open **My Decimal Setup.**

8. Select the **11 X 17 (1 to 1)** tab.

9. Erase the Viewport frame. (Click on it, select erase and <enter>)

10. Select the Viewport layer.

Note: <u>Lt users can not convert objects to viewports</u>. But you can still do this exercise. Instead of drawing circles, draw two single viewports approximately the size of the circles shown below. Then skip to instruction 12.

11. Create 2 new Viewports as shown below as follows:
    a. Draw 2 circles, **R3.50**, as shown below.
    b. Select the "**Convert Object to Viewport**" icon from the Viewport toolbar.
    c. Select the each new Circle. (they are now viewports)

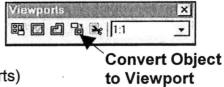

Convert Object to Viewport

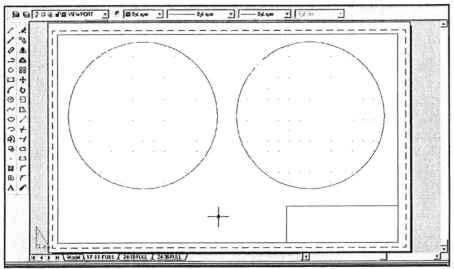

12. **Add Viewport Titles** as shown below. (**In paperspace**)
    a. Use "Single Line Text"
    b. Style = Class Text
    c. Height = .35
    d. Layer = Text Heavy

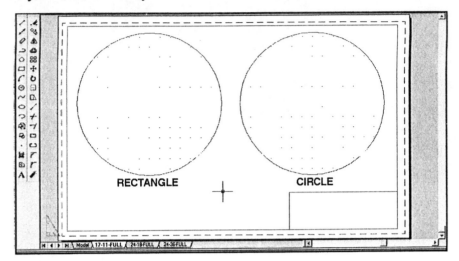

13. Double click inside of the Rectangle Viewport (on the left) to **activate Model space**.

14. Change to the XREF layer.

15. **Xref** the **Rect** drawing as follows:
    a. Select **INSERT / EXTERNAL REFERENCE**
    b. Find and Select **Rect** drawing
    c. Select the **OPEN** button.
    d. Select the **OK** button.
    e. Click inside the "Rectangle" Viewport.

16. Use **Zoom / All** inside each viewport so you can see the rectangle.

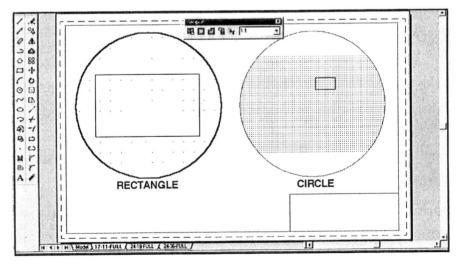

17. **Adjust the scale** in the **Rectangle** Viewport to **1 : 1.**

18. **"Pan"** the **Rect** drawing inside of the Rectangle viewport to display it as shown. *(**Do not use the Zoom commands or you will have to re-adjust the scale**)*

11-18

Now we are going to make the Rect drawing disappear in the Circle viewport.

19. Freeze all the layers that belong to **Rect** drawing in the **Circle Viewport** as follows:
    a. **Activate** the Circle Viewport
    b. Select **FORMAT / LAYER**
    c. Select **Rect** in the **Named layer filters**
    d. Select all the layers that begin with **Rect.**

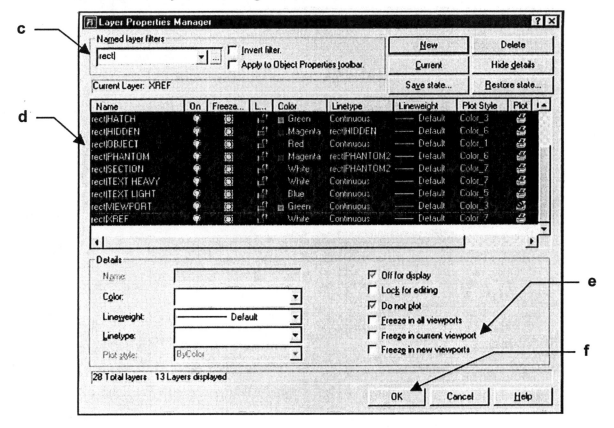

e. Select the **"Freeze in current Viewport"** box.
f. Select **OK** button.

The Rect drawing, inside the Circle Viewport, should have disappeared.

20. **Xref** drawing **Circle** into the **Circle Viewport** as follows:
    a. Select the **XREF** layer.
    b. Select **INSERT / EXTERNAL REFERENCE**
    c. Find and Select the **Circle** drawing.
    d. Select the **OPEN** button.
    e. Select the **OK** button.
    f. Click inside the Circle Viewport.

21. Use **Zoom / All** inside the Circle viewport if you can't see the Circle drawing.

22. **Adjust the scale** in the **Circle** Viewport to **1 : 1.**

23. **"Pan"** the **Circle** drawing inside of the Circle viewport to display it as shown.
    (Do not use the Zoom commands or you will have to re-adjust the scale)

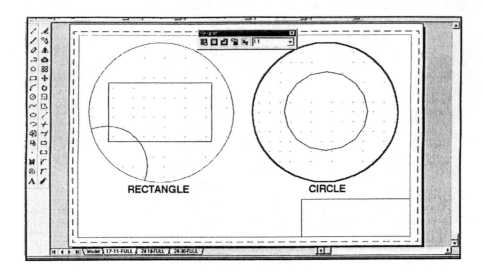

Now we will make the Circle drawing disappear in the Rectangle viewport.

24. Freeze all the layers, that belong to the Circle drawing, in the Rectangle Viewport as follows:
    a. **Activate** the Rectangle Viewport
    b. Select **FORMAT / LAYER.**
    c. Select the **Circle** in the **Named layer filters** box
    d. Select all the layers that begin with **Circle**.
    e. Select the "**Freeze in current Viewport**" box.
    f. Select **OK** button.

The Xref drawing, Circle, should have disappeared in the Rectangle Viewport.

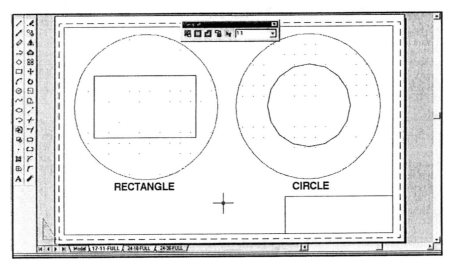

25. **Change to Paper space and change the Title Block**
    a. Title = Multiple Xref drawings
    b. Scale = 1 - 1

26. Save as **EX-11A**

27. **Plot** using Page Set up **11 X 17 (1 to 1) All Black.**

EXERCISE 11B

Creating Multi-scaled Views

1. Open **My Decimal Setup.**
2. Select the **11 X 17 (1 to 1)** tab.
3. Important: Erase the existing Viewport frame. (Click on it, select erase and <enter>)
4. Select the Viewport layer.
5. Create 4 new Viewports as shown below.

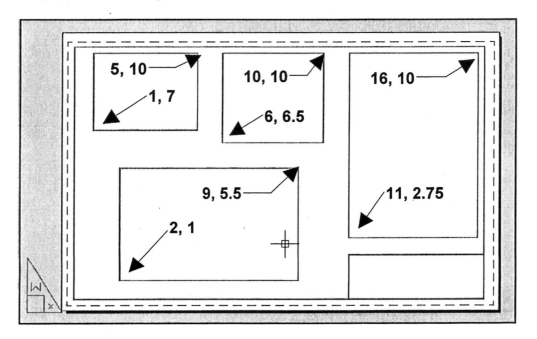

6. Add Viewport Titles as shown below. (In paperspace)
   a. Use "Single Line Text"
   b. Style = Class Text
   c. Height = .35
   d. Layer = Text Heavy

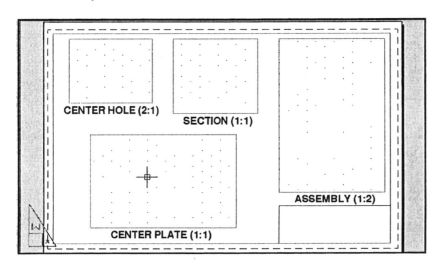

7. Double click inside one of the Viewports to **activate Model space**.

8. **Xref** drawing 6A as follows:
   a. Select **INSERT / EXTERNAL REFERENCE**
   b. Find and Select 6A
   c. Select the OPEN button.
   d. Remove the check mark from all of the "Specify On-Screen" boxes.
   e. Select the OK button.

9. **Adjust the scale** in each Viewport to the scale listed under each viewport as shown below.

10. **"Pan"** the drawing inside each viewport to display the correct area. (Do not use the Zoom commands or you will have to re-adjust the scale)

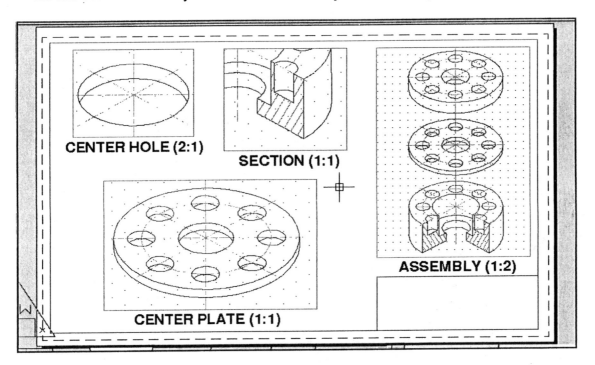

11. Set **"PSLTSCALE"** command to **"0"** and **Regenall.** (Refer to page 11-15)

12. **Return to Paper space and change the Title Block**
   a. Title = Multi-scaled Viewports
   b. Scale = Noted

13. Save as **EX-11B**

14. **Plot** using Page Set up **11 X 17 (1 to 1) All Black.**

EXERCISE 11C

Unload, Reload and Detach an Xref drawing.

1. Open **EX-11B**.
2. Select the **11 X 17 (1 to 1)** tab.
3. Select **INSERT / XREF MANAGER**

The following dialog box should appear.

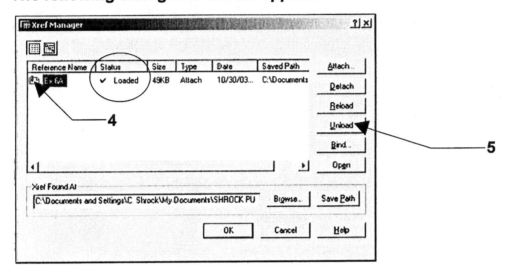

4. Select the xref drawing **EX-6A** from the Reference Name list.

5. Select the **Unload** button and then the **OK** button.

   Did the drawing disappear?

6. Select **INSERT / XREF MANAGER** again.

   Notice EX-6A is still listed.

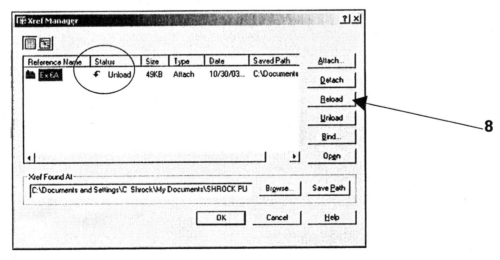

7. Select the **EX-6A** drawing from the Reference Name list again.

8. Select the **Reload** button and then the **OK** button.

   Did the drawing re-appear?

9. Select **INSERT / XREF MANAGER** again.

10. Select the EX-6A drawing from the Reference Name List again.

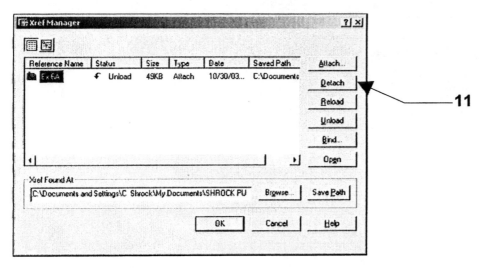

11. Select the **DETACH** button and then the **OK** button.

   Did the drawing disappear?

12. Select **INSERT / XREF MANAGER** <u>one more time</u>.

   This time, EX-6A should no longer appear in the "Reference Name" list.

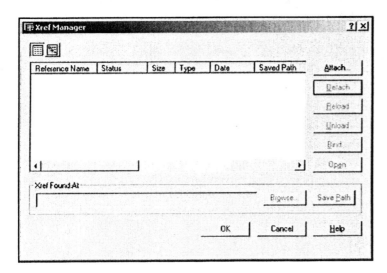

13. **Save as EX-11C. <u>Do not Plot</u>.**

EXERCISE 11D

CLIPPING AN EXTERNAL REFERENCE

1. Open **EX-11B.**
2. Select the **Model** tab.
3. Select **View / Zoom / Extents**

4. Select **XCLIP** command. (Ref to page 11-8)
5. Clip the lower section to appear as **shown below**:
   a. Select <u>New Boundary</u> and <u>Rectangular</u>

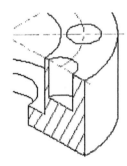

6. **Save as EX-11D1**

7. Select the **XCLIP** command <u>again</u>.
   a. Select **OFF**
   The entire drawing should reappear.

8. Select the **XCLIP** command <u>again</u>.
9. Clip the lower section to appear approximately as shown **below left**:
   a. Use <u>New Boundary</u> and <u>Polygonal</u>.
   b. Select "<u>Yes</u>" when prompted to "<u>Delete old boundary</u>"
10. **Save as EX-11D2**

11. Select the **XCLIP** command <u>again</u>.
12. Generate a visible Polyline boundary as shown **below right**.
    a. Use <u>Generate Polyline</u>.
13. **Save as EX-11D3**

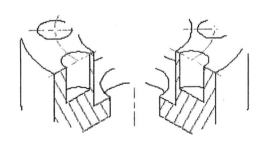
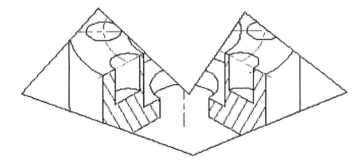

NOTES:

LEARNING OBJECTIVES

After completing this lesson, you will be able to:

1. Dimension using Datum dimensioning
2. Use alternate dimensioning for inches and Millimeters
3. Assign Tolerances to a part
4. Use Geometric tolerances
5. Typing Geometric Symbols

LESSON 12

ORDINATE Dimensioning

Ordinate dimensioning is primarily used by the sheet metal industry. But many others are realizing the speed and tidiness this dimensioning process allows.

Ordinate dimensioning is used when the X and the Y coordinates, from one location, are the only dimensions necessary. Usually the part has a uniform thickness, such as a flat plate with holes drilled into it. The dimensions to each feature, such as a hole, originate from one "datum" location. This is similar to "baseline" dimensioning. Ordinate dimensions have only one datum. The datum location is usually the lower left corner of the object.

Ordinate dimensions appearance is also different. Each **dimension** has only one **leader line** and a **numerical value**. Ordinate dimensions do not have extension lines or arrows.

Example of Ordinate dimensioning:

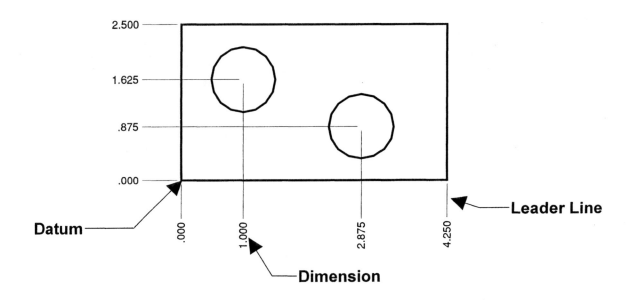

Note:
Ordinate dimensions can be Associative and are Trans-spatial. Which means that you can dimension in paperspace and the ordinate dimensions will remain associated to the object they dimension. (Except for Qdim ordinate)

Refer to the next page for step by step instructions to create Ordinate dimensions.

Creating Ordinate dimensions

1. Move the "Origin" to the desired "datum" location as follows:
   a. Select **TOOLS / MOVE UCS**
   b. Snap to the desired location.
2. Select the Ordinate command using one of the following:

   TYPE = DIMORDINATE
   PULLDOWN = DIMENSION / ORDINATE
   TOOLBAR = DIMENSION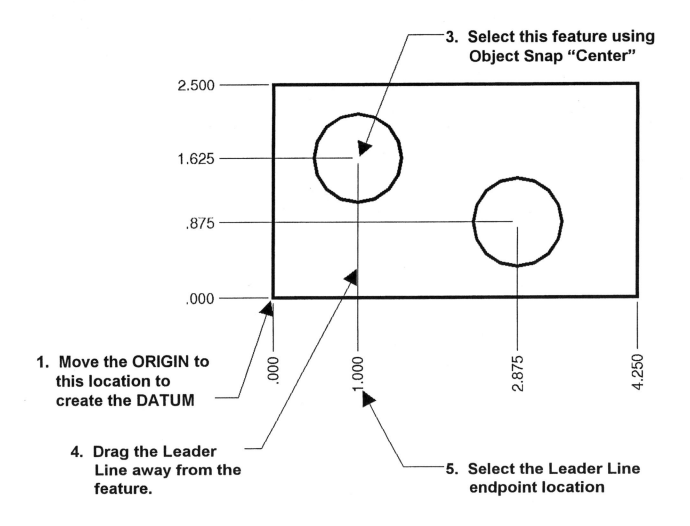

3. Select the first feature, using object snap.
4. Drag the leader line horizontally or vertically away from the feature.
5. Select the location of the "leader endpoint.
   (The dimension text will align with the leader line)

Use "Ortho" to keep the leader lines straight.

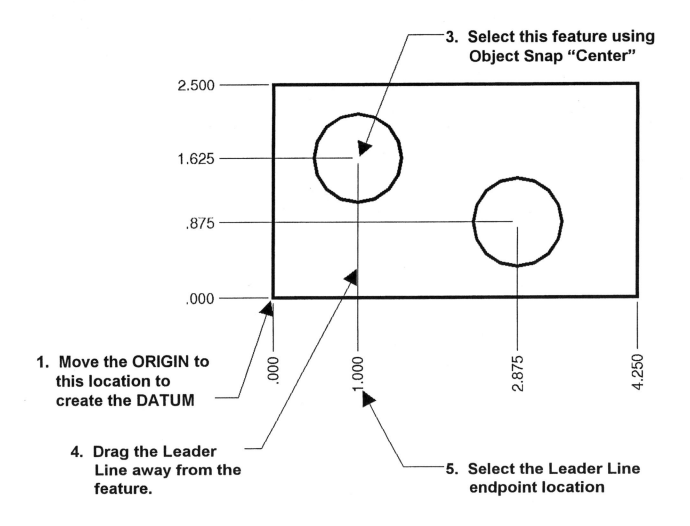

3. Select this feature using Object Snap "Center"

2.500

1.625

.875

.000

.000

1.000

2.875

4.250

1. Move the ORIGIN to this location to create the DATUM

4. Drag the Leader Line away from the feature.

5. Select the Leader Line endpoint location

JOG an Ordinate dimension

If there is insufficient room for a dimension you may want to jog the dimension.
To **"jog"** the dimension, as shown below, turn "**Ortho**" **off** before placing the Leader Line endpoint location. The leader line will automatically jog. With Ortho off, you can only indicate the feature location and the leader line endpoint location, the leader line will jog the way it wants to.

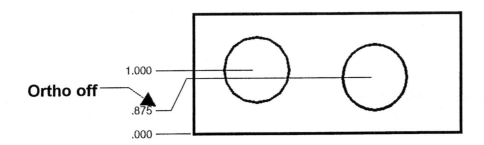

Qdim with Ordinate dimensioning

1. Select **DIMENSION / QDIM** (Refer to Exercise Workbook for Beginning Acad, lesson 20)
2. Select the geometry to dimension <enter>
3. Type **"O"** <enter> to select Ordinate
4. Type **"P"** <enter> to select the **datumPoint** option.
5. Select the datum location on the object. (use Object snap)
6. Drag the dimensions to the desired distance away from the object.

> **Note:** Qdim can be associative but is not trans-spatial.
> **If the object is in Model Space, you must dimension in Model Space.**

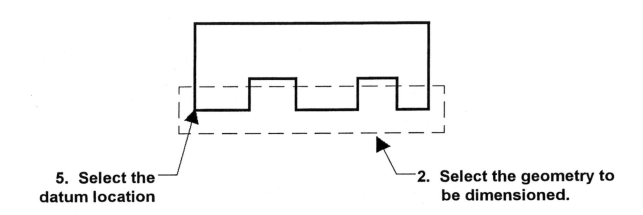

5. Select the datum location

2. Select the geometry to be dimensioned.

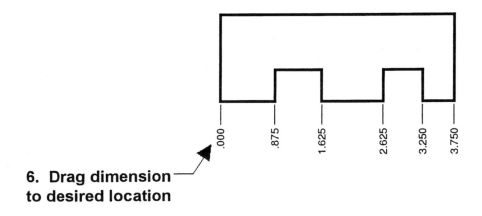

6. Drag dimension to desired location

.000   .875   1.625   2.625   3.250   3.750

ALTERNATE UNITS

The options in this tab allow you to display inches as the primary units and the millimeter equivalent as alternate units. The millimeter value will be displayed inside brackets immediately following the inch dimension. Example: 1.00 [25.40]

1. Select **DIMENSION / STYLE / MODIFY**
2. . Select the **ALTERNATE UNITS** tab.

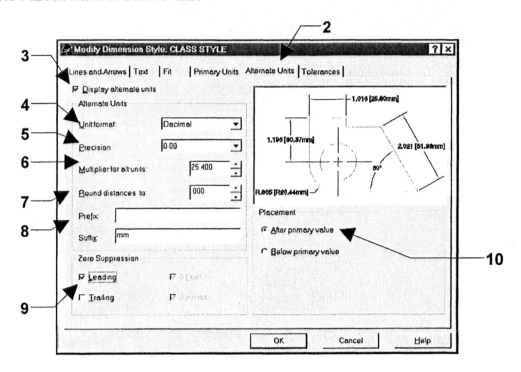

3. **Display alternate units.** Check this box to turn ON alternate units

4. **Unit format.** Select the Units for the alternate units.

5. **Precision** Select the Precision of the alternate units. This is independent of the Primary Units.

6. **Multiplier for all units** The primary units will be multiplied by this number to display the alternate unit value.

7. **Round distance to** Enter the desired increment to round off the alternate units value.

8. **Prefix / Suffix** This allows you to include a Prefix or Suffix to the alternate units. Such as: type **mm** to the Suffix box to display **mm** (for millimeters) after the alternate units.

9. **Zero Suppression** If you check one or both of these boxes, it means that the zero will not be drawn. It will be suppressed.

10. **Placement** Select the desired placement of the alternate units. Do you want them to follow immediately after the Primary units or do you want the Alternate units to be below the primary units?

TOLERANCES

When you design and dimension a widget, it would be nice if when that widget was made, all of the dimensions were exactly as you had asked. But in reality this is very difficult and or expensive. So you have to decide what actual dimensions you could live with. Could the widget be just a little bit bigger or smaller and still work? This is why tolerances are used.

A **Tolerance** is a way to communicate, to the person making the widget, how much larger or smaller this widget can be and still be acceptable. In other words each dimension can be given a maximum and minimum size. But the widget must stay within that **"tolerance"** to be correct. For example: a hole that is dimensioned 1.00 +.06 -.00 means the hole is nominally 1.00 but it can be as large as 1.06 but can not be smaller than 1.00.

1. Select **DIMENSION / STYLE / MODIFY**
2. Select the **TOLERANCES UNITS** tab.

> *Note: if the dimensions in the display look strange,*
> *make sure "Alternate Units" are turned OFF.*

3. **Method**
The options allows you to select how you would like the tolerances displayed. There are 5 methods: None, Symmetrical, Deviation, Limits. (Basic is used in geometric tolerancing and will not be discussed at this time)

Refer to the next page for descriptions of methods.

4. **Scaling for height**. This controls the height of the tolerance text. The entered value is a percentage of the primary text height. If .50 is entered, the tolerance text height will be 50% of the primary text height.

5. **Vertical position**. This controls the placement of the tolerance text in relation to the primary text. The options are Top, Middle and Bottom. Whichever option you select, it will align the tolerance text with the bottom of the primary text.

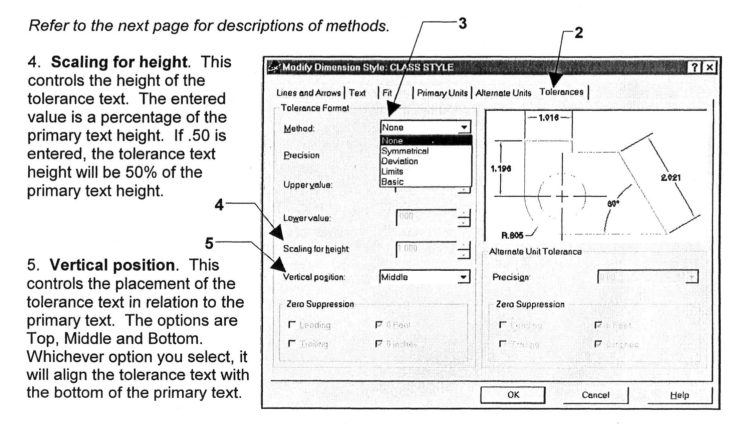

SYMMETRICAL is an equal bilateral tolerance. It can vary as much in the plus as in the negative. Because it is equal in the plus and minus direction, only the "Upper value" box is used. The "Lower value" box is grayed out.

Example of a Symmetrical tolerance :

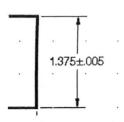

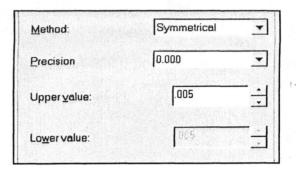

DEVIATION is an unequal bilateral tolerance. The variation in size can be different in both the plus and minus directions. Because it is different in the plus and the minus the "Upper" and "Lower" value boxes can be used.

Example of a Deviation tolerance:

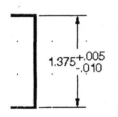

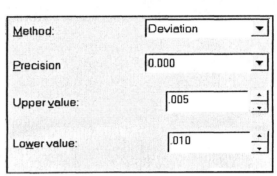

Note: If you set the upper and lower values the same, the tolerance will be displayed as symmetrical.

LIMITS is the same as deviation except in how the tolerance is displayed. Limits calculates the plus and minus by adding and subtracting the tolerances from the nominal dimension and displays the results. Some companies prefer this method because no math is necessary when making the widget. Both "Upper and Lower" value boxes can be used.
Note: The "Scaling for height" should be set to "1".

Example of a Limits tolerance:

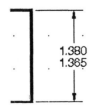

GEOMETRIC TOLERANCING

Geometric tolerancing is a general term that refers to tolerances used to control the form, profile, orientation, runout, and location of features on an object. Geometric tolerancing is primarily used for mechanical design and manufacturing. The instructions below will cover the Tolerance command for creating geometric tolerancing symbols and feature control frames.

If you are not familiar with geometric tolerancing, you may choose to skip this lesson.

1. Select the **TOLERANCE** command using one of the following:

    TYPE = TOL
    PULLDOWN = DIMENSION / TOLERANCE
    TOOLBAR = DIMENSION

    The Geometric Tolerance dialog box, shown below, should appear.

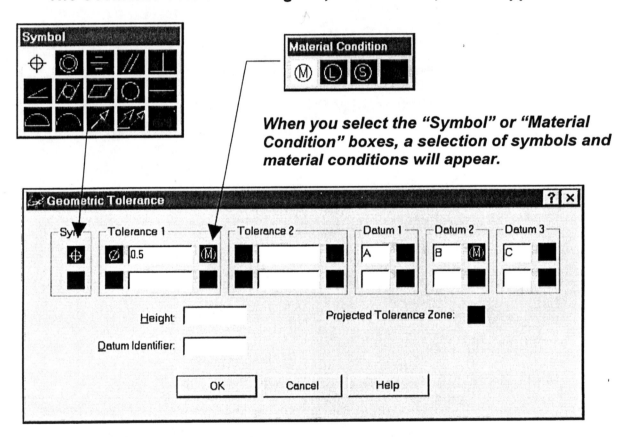

When you select the "Symbol" or "Material Condition" boxes, a selection of symbols and material conditions will appear.

2. Make your selections and fill in the tolerance and datum boxes.
3. Select the **OK** box
4. The tolerance should appear attached to your cursor. Move the cursor to the desired location and press the left mouse button.

Note: the size of the Feature Control Frame above, is determined by the height of the dimension text.

GEOMETRIC TOLERANCES and QLEADER

The **Qleader** command allows you to draw leader lines and access the dialog boxes used to create feature control frames in one operation.

1. Select **Dimension / Leader**
2. Select the **"Settings"** option. (Right click, and select Settings from the short cut menu)

   The Leader Settings dialog box should appear.

3. Select the Annotation tab.
4. Select Tolerance then OK.

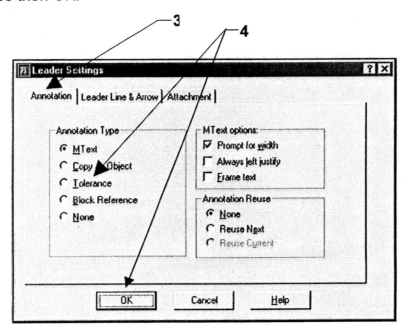

5. Place the first leader point. **P1**
6. Place the next point **P2**
7. Press <enter>

The Geometric Tolerance dialog box will appear.

8. Make your selections and fill in the tolerance and datum boxes.

9. Select the **OK** button.

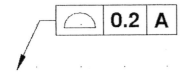

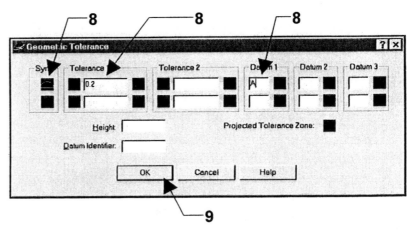

12-10

DATUM FEATURE SYMBOL

A datum in a drawing is identified by a **"datum feature symbol"**.

To create a *datum feature symbol*:

1. Select Dimension / Tolerance
2. Type the "datum reference letter" in the "Datum Identifier" box.
3. Select the OK button.

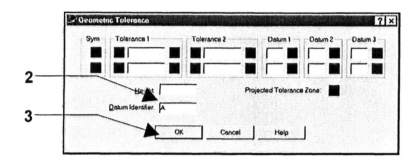

To create a *datum feature symbol* combined with a *feature control frame*:

1. Select **Dimension / Tolerance**
2. Make your selections and fill in the tolerance.
3. Type the "datum reference letter" in the "Datum Identifier" box.
4. Select the **OK** button.

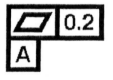

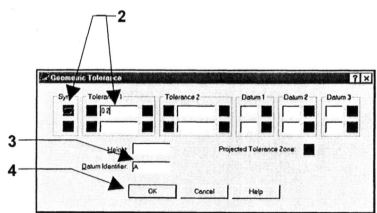

DATUM TRIANGLE

A datum feature symbol, in accordance with ASME Y14.5M-1994, includes a leader line and a datum triangle filled. You can create a wblock or you can use the two step method below using Dimension / Tolerance and Qleader.

1. Select **Dimension / Leader**.
2. Select the **"Settings"** option. (Right click, and select Settings from the short cut menu)

The Leader Settings dialog box should appear.

3. Select the Annotation tab.	5. Select the Leader Line & Arrow tab.
4. Select the Tolerance button.	6. Select Datum Triangle Filled
	7. Select the OK button.

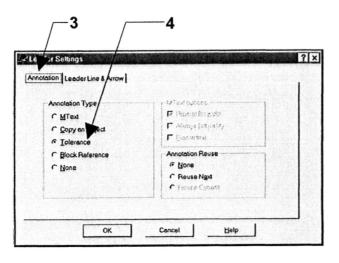

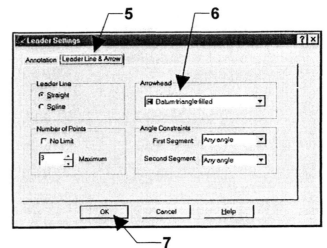

8. Specify the first leader point. (The triangle endpoint)
9. Specify next point then press <enter>
10. When the Geometric Tolerance dialog box appears, select the OK button.

If you were successful, a __datum triangle filled__ with a __leader line__ should appear. (As shown below)

11. Next create a datum feature symbol. (Follow the instructions on the previous page.)

12. Now move the datum feature symbol to the endpoint of the leader line to create the symbol below left.

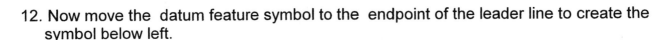

You are probably wondering why we didn't just type "A" in the identifier box. That method will work if your leader line is horizontal. But if the leader line is vertical, as shown on the left, it will not work. (The example on the right illustrates how it would appear)

TYPING GEOMETRIC SYMBOLS

If you want geometric symbols in the notes that you place on the drawing, you can easily accomplish this using a font named **GDT.SHX**. This font will allow you to type normal letters and geometric symbols, in the same sentence, by merely toggling the SHIFT key up and down with CAPS LOCK on.

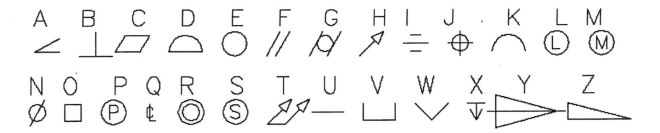

1. First you must create a new text style using the **GDT.SHX** font.

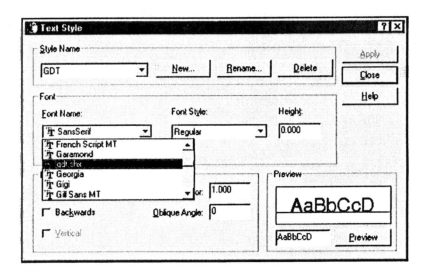

2. **CAPS LOCK** must be **ON**.

3. Select **DRAW / TEXT / Single Line or Multiline**

4. Now type the the sentence shown below. When you want to type a symbol, press the **SHIFT** key and type the letter that corresponds to the symbol. For example: If you want the diameter symbol, press the shift key and the "N" key. (Refer to the alphabet of letters and symbols shown above.)

$$3X \quad \varnothing.44 \quad \sqcup\!\varnothing1.06 \quad \overline{\vee}.06$$

Can you decipher what it says?
(Drill (3) .44 diameter holes with a 1.06 counterbore .06 deep)

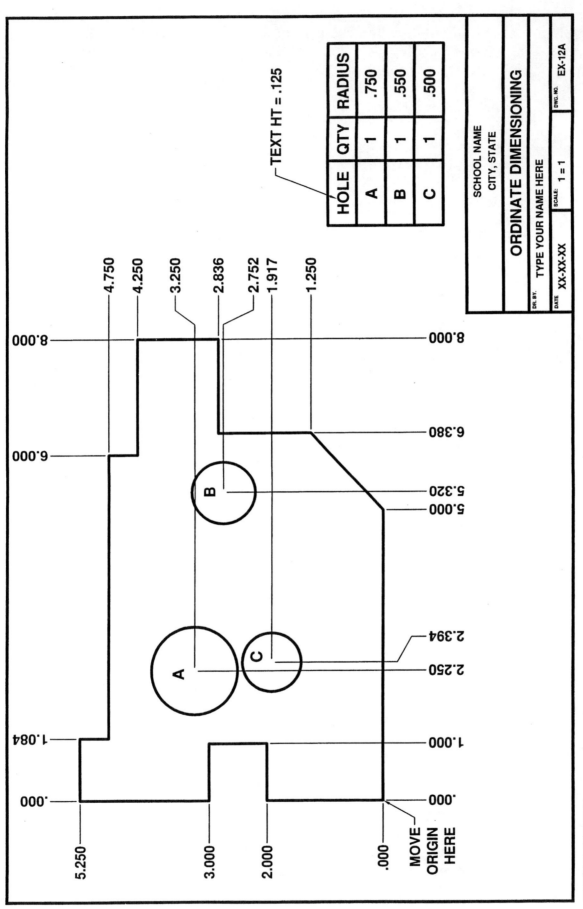

TEXT HT = .125

HOLE	QTY	RADIUS
A	1	.750
B	1	.550
C	1	.500

SCHOOL NAME
CITY, STATE

ORDINATE DIMENSIONING

DR. BY.   TYPE YOUR NAME HERE

DWG. NO.   EX-12A

SCALE: 1 = 1

DATE   XX-XX-XX

EXERCISE 12A

INSTRUCTIONS:

1. Open MY DECIMAL SETUP and select the 11 X 17 (1 to 1) tab.
2. Draw the drawing above inside the viewport, in model space.
3. Dimension using ORDINATE dimensioning (Refer to pages 12-2 thru 5)
4. Make sure the model space scale is 1:1.
5. Save as: EX-12A
6. Plot using "11 X 17 (1 to 1) All Black" Page Setup.

MOVE
ORIGIN
HERE

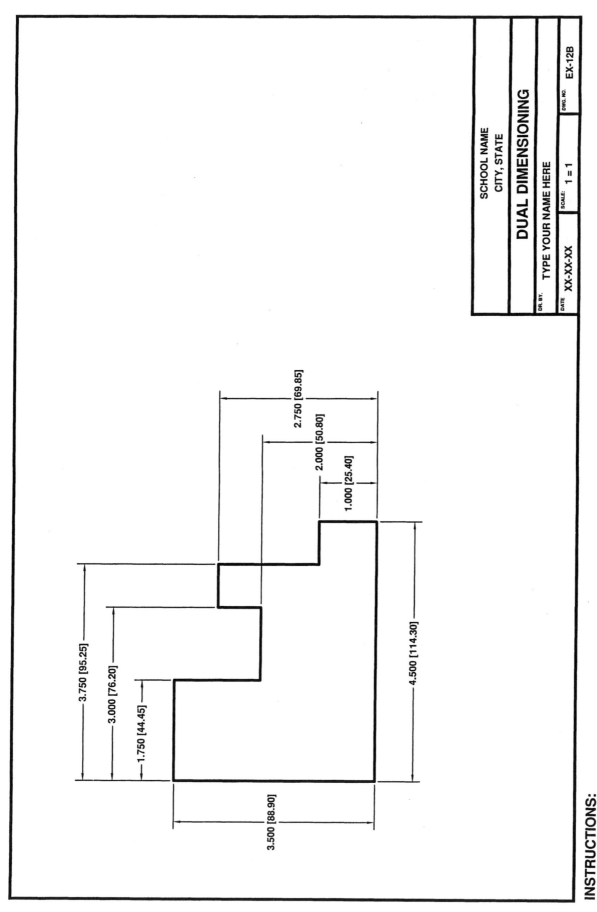

DUAL DIMENSIONING

SCHOOL NAME
CITY, STATE

DR. BY. TYPE YOUR NAME HERE

DATE XX-XX-XX | SCALE: 1 = 1 | DWG. NO. EX-12B

EXERCISE 12B

2.750 [69.85]

2.000 [50.80]

1.000 [25.40]

3.750 [95.25]

3.000 [76.20]

1.750 [44.45]

4.500 [114.30]

3.500 [88.90]

INSTRUCTIONS:

1. Open MY DECIMAL SETUP and select the 11 X 17 (1 to 1) tab.
2. Draw the drawing above inside the viewport, in model space.
3. Dimension using ALTERNATE UNITS (Refer to Page 12-6)
4. Make sure the model space scale is 1:1.
5. Save as: EX-12B
6. Plot using "11 X 17 (1 to 1) All Black" Page Setup.

12-15

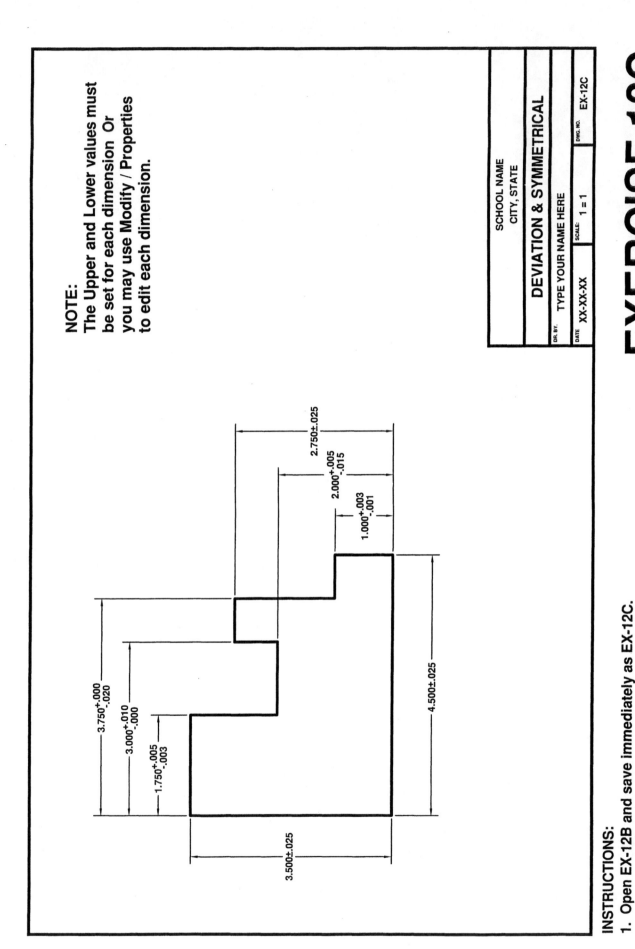

NOTE:
The Upper and Lower values must
be set for each dimension Or
you may use Modify / Properties
to edit each dimension.

SCHOOL NAME
CITY, STATE

DEVIATION & SYMMETRICAL

DR. BY.   TYPE YOUR NAME HERE

DATE XX-XX-XX   SCALE: 1 = 1   DWG. NO.   EX-12C

EXERCISE 12C

3.750⁺·⁰⁰⁰ / ₋·⁰²⁰

3.000⁺·⁰¹⁰ / ₋·⁰⁰⁰

1.750⁺·⁰⁰⁵ / ₋·⁰⁰³

2.750±.025

2.000⁺·⁰⁰⁵ / ₋·⁰¹⁵

1.000⁺·⁰⁰³ / ₋·⁰⁰¹

4.500±.025

3.500±.025

INSTRUCTIONS:
1. Open EX-12B and save immediately as EX-12C.
2. Change the Dimensions to DEVIATION & SYMMETRICAL (Refer to Page 12-7 & 8)
3. You may erase the dimension or use Modify/Properties.
4. Save as: EX-12C
5. Plot using "11 X 17 (1 to 1) All Black" Page Setup.

12-16

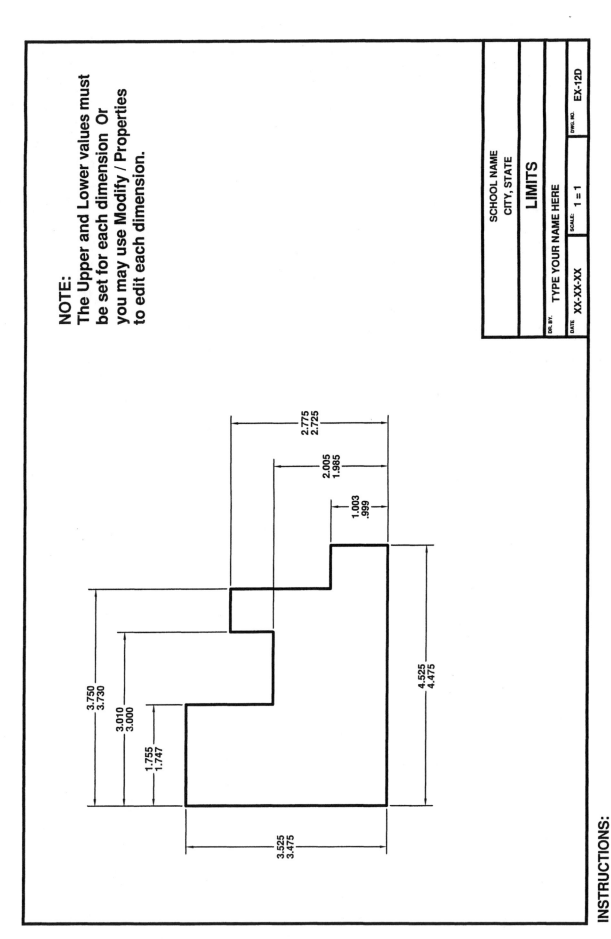

NOTE:
The Upper and Lower values must be set for each dimension Or you may use Modify / Properties to edit each dimension.

SCHOOL NAME
CITY, STATE

LIMITS

DR. BY.   TYPE YOUR NAME HERE

DATE   XX-XX-XX   SCALE: 1 = 1   DWG. NO.   EX-12D

EXERCISE 12D

INSTRUCTIONS:

1. Open EX-12B or EX-12C and save immediately as EX-12D.
2. Change the Dimensions to LIMITS (Refer to Page 12-7 & 8)
3. You may erase the dimension or use Modify/Properties.
4. Save as: EX-12D
5. Plot using "11 X 17 (1 to 1) All Black" Page Setup.

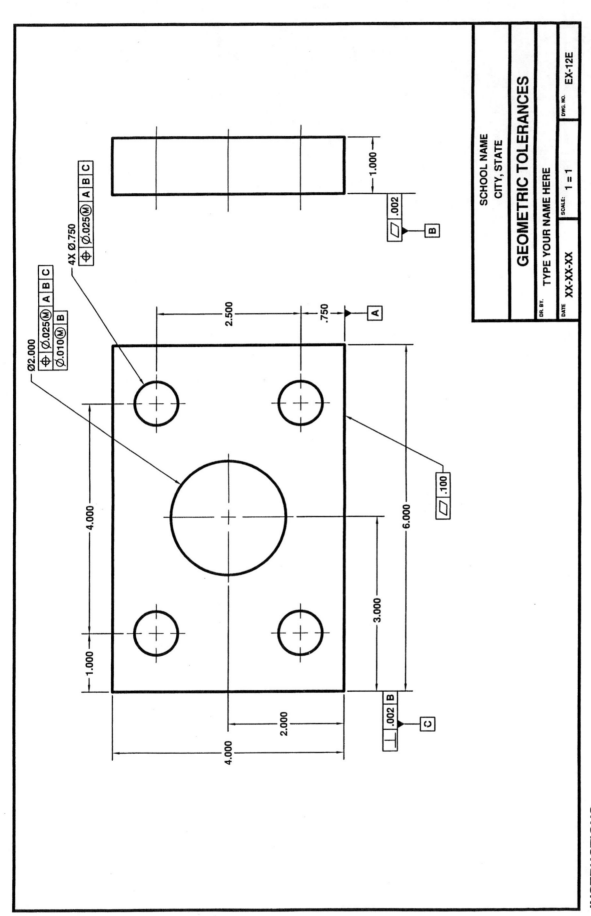

EXERCISE 12E

INSTRUCTIONS:

1. Open MY DECIMAL SETUP and select the 11 X 17 (1 to 1) tab.
2. Draw the drawing above inside the viewport, in model space.
3. Dimension using GEOMETRIC TOLERANCES (Refer to Page 12-9 thru 12)
4. Make sure the model space scale is 1:1.
5. Save as: EX-12E
6. Plot using "11 X 17 (1 to 1) All Black" Page Setup.

SCHOOL NAME
CITY, STATE

GEOMETRIC TOLERANCES

DR. BY.   TYPE YOUR NAME HERE

SCALE: 1 = 1

DWG. NO.   EX-12E

DATE   XX-XX-XX

LEARNING OBJECTIVES

After completing this lesson, you will be able to:

1. Understand the concept of 3D
2. Change the display as a hidden view
3. Understand the difference between:
   Wireframe, Surface and Solid Modeling

The following lessons are an introduction to 3D. These lessons will give you a good understanding of the basics of AutoCAD's 3D program.

Unfortunately, LT users cannot create solids.

LESSON 13

INTRODUCTION TO 3D

We live in a three dimensional world, yet most of our drawings represent only two dimensions. In Lesson 6, you made an isometric drawing that appeared three dimensional. In reality, it was merely two dimensional lines drawn on angles to give the appearance of depth.

In the following lessons you will be presented with a basic introduction to AutoCAD's 3D techniques for constructing and manipulating objects. These lessons are designed to give you a good start into the environment of 3D and encourage you to continue your education in the world of CAD.

DIFFERENCES BETWEEN 2D AND 3D

Axes
In 2D drawings you see only one plane. This plane has two axes, X for horizontal and Y for vertical. In 3D drawings an additional plane is added. This third plane is defined by an additional axis called Z. The direction of the positive Z axis basically comes out of the screen toward you and gives objects height. To draw a 3D object you must input all three, X, Y and Z axes coordinates.

An simple way to visualize these axes is to consider the X and Y axes as the ground and the Z axes as a Tree growing up (positive coordinates) from the ground or the roots growing down (negative coordinates) into the ground.

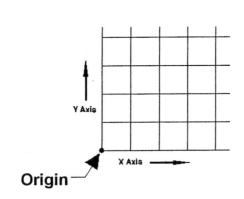

2D Coordinate System

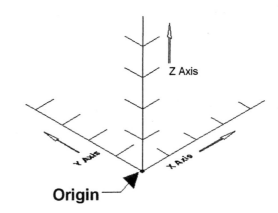

3D Axes Coordinate System

BASIC TYPES OF 3D MODELS

There are 3 basic types of 3D models.

1. Wireframe models
2. Surface models
3. Solid models.

A brief description of each starts on page 13-6.

A Peek into the 3D World

Just to get you a little excited about 3D, let's experiment a little with the View and Shade options. This will give you a basic understanding of the difference between a 2D and 3D viewing.

Step 1. Open the drawing "**3D Helper.dwg**".
(This is one of the drawings you downloaded from Shrock Publishing website.)

Notice the Icon ——

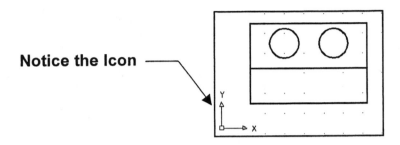

This is a view of the X and Y plane (the ground). This should appear very familiar to you.

Step 2. Select **View / 3D Views / SE Isometric**

Notice the Icon changed ——

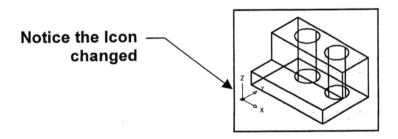

The object is now considered a "**3D Model**" not a "**2D Object**".
You are viewing the X, Y and Z axes (height). The object is shown as a "2D Wireframe".)
(You are now viewing the **SE (South East) Isometric** view of the "**Model**".

Step 3. Select **View / Shade / Gouraud Shaded** or

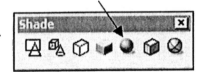

Notice the Icon changed again ——

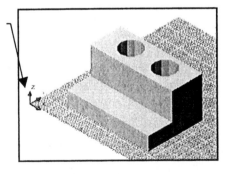

Select the 2D or 3D wireframe icon to return to a Wireframe display

Now is when the fun starts!

We will cover all of this and more in the Lessons that follow.

Viewing a 3D Model

It is very important for you to be able to control how you view the 3D Model. The process of changing the view is called **selecting the Viewpoint**. The **Viewpoint** is the location where **"YOU"** are standing. ***This is a very important concept to understand***. The Model does not actually move, you move. For example, if you want to look at the South East corner of your house, you need to walk to the South East corner of your lot and look at the corner of your house. Your house did not move, but you are seeing the South East corner of your house. If you want to see the Top or Plan View of your house, you would have to climb up on the roof and look down on the house. The house did not move, you did.

As you select the Viewpoints on the toolbar below, remember, you are selecting where you are standing. The view that appears on the screen is what you would see when you move to the viewpoint and looked at the model.

VIEW TOOLBAR

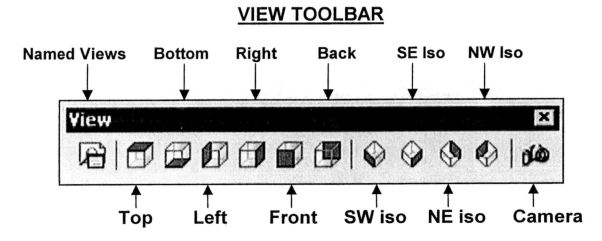

Note: You may also use the Pulldown menu **VIEW / 3D VIEWS**
Although the toolbars are more handy, sometimes they clutter up your drawing area.
(It is your preference.)

3D Orbit
3D Orbit allows you to dynamically rotate the viewpoint of the 3D Model. (walk around it)

TYPE = 3do
PULLDOWN = VIEW / 3D ORBIT

TOOLBAR = 3D ORBIT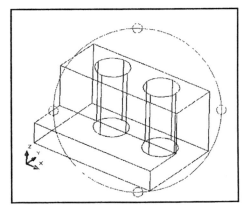

A green circle appears in the middle of the current viewport. This is called the Trackball. Place the cursor inside or outside of the trackball, click and drag the cursor. The view is dynamically changed as you drag the cursor. When done, right click and select Exit or press the ESC key.

To return to an isometric view, select one of the 3D Views mentioned above.

Note: This command may be used in the middle of any drawing command.

Hiding Lines

You may have noticed when you displayed the model, it displayed all of the lines including the lines that represent the internal features. This called "**Wireframe Display**". This is not the same as "Wireframe Model" as described on page 13-6. The model is still a solid. It is merely displayed with all of the lines visible to make it easier to select the lines.

Sometimes this gets somewhat confusing and you would like the view to be displayed without the internal lines (hidden) displayed. To create a "**Hidden Display**", in the current viewport, you must use the Hide command.

1. First select the **Viewport** that you want to Hide.
   (Note: If you have multiple viewports, you must use the hide command in each viewport)

2. Select the Hide Command using one of the following:

   TYPE = HI
   PULLDOWN = VIEW / HIDE

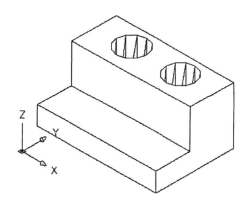

Dispsilh variable = 0

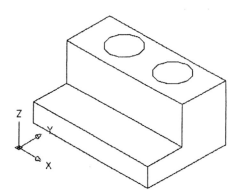

Dispsilh variable = 1

3. To return to a wireframe display, select the 2D or 3D icon.

DISPSILH variable

Notice that the Hide command uses zigzag lines, called facets, to indicate internal surfaces. The appearance of the facets is controlled by the **dispsilh** variable.
(dispsilh = Display Silhouette)
If you do not want this appearance, you use the **dispsilh** variable to display 3D solid objects with silhouette **edges** only.

1. Command: *type: dispsilh <enter>*
2. Enter new value for DISPSILH <0>: *type 1 <enter>*
3. Command: *type: hi <enter>*

WIREFRAME MODEL

A wireframe model of a box is basically 12 pieces of wire (lines). Each wire represents an **edge** of the object. You can see through the object because there are no surfaces to obscure your view. This type of model does not aid in the visualization of the 3D object. Wireframe models have no volume or mass. The hidden line removal command cannot be used on a wireframe model.

How to draw a Wireframe Box

1. Draw a **Rectangle**

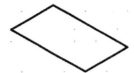

2. Select **VIEW / 3D / SE Isometric** or

3. Copy the Rectangle 5" **above** the original rectangle as follows:
   a. Select the Copy command
   b. Select the Rectangle then <enter>
   c. Select one of the corners as the **basepoint**
   d. Type the **X, Y, Z** coordinates for the new location: **@ 0, 0, 5**

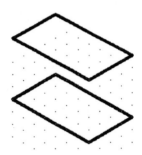

*Now think about the coordinates entered. The "@" sign of course is very important because you want the new rectangle location to be "relative" to the existing rectangle. The coordinates **0, 0, 5** mean that you **do not** want the new rectangle location to move in the **X** or **Y** axis. But you **do** want the new rectangle location to move **5" in the positive "Z"** axis.*

4. Now draw lines from the corners of the one rectangle to the other rectangle.

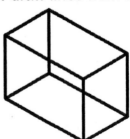

That is all there is to it. You have now completed a Wireframe Box.
This Wireframe can now be used as the structure for a Surface Model described on the next page.

SURFACE MODELS

A Surface Model is like an empty cardboard box. All surfaces and edges are defined but there is nothing inside. The model appears to be solid but is actually an empty shell. A surface model makes a good visual <u>representation</u> of a 3D object. Because the front surfaces obscure the back surfaces from view, the hidden line removal command can be used.

You may attach a 3D Surface to a Wireframe structure or you may use one of AutoCAD's pre-defined 3D Objects. (Refer to Lesson 10 for more discussion on Surface Modeling)

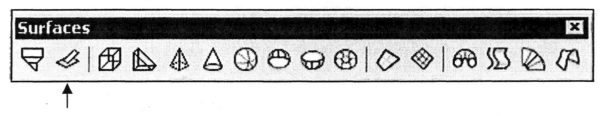

How to add a 3D Surface to a Wireframe structure

1. Create a Wireframe Model as shown on page 13-6.

2. Select **VIEW / 3D / SE Isometric** or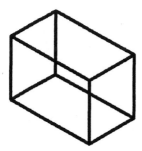

3. Select **DRAW / SURFACES / 3D FACE** or

4. Draw a (2) 3D Surface by snapping to the 4 corners as shown
    below and **\<enter\>** (the Shape will **Close** automatically.)
    Note: Each surface must be created as a separate object.

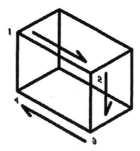

<u>*I know that the surface does not seem to be there but it is. The next step will make it appear.*</u>

5. Select **View / Hide** or Type: **hi** or

Your 3D Wireframe model now has a surface attached.
You may even use the Shade command

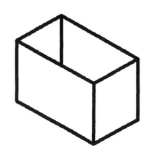
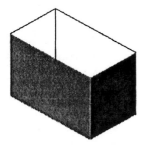

SOLID MODELS *(Refer to Lesson 2 for detailed instructions)*

*It is important that you understand both Wireframe and Surface Modeling **but** you will not be using either very often. Solid Modeling is the most useful and the most fun.*

Solid models have edges, surfaces and **mass**.
You can display the solid model as a wireframe display or hidden display.

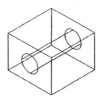 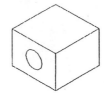

Solid Models can be modified using many **Solid Editing** features such as the ones described below. *(Refer to Lessons 3 and 4 for more detailed instructions)*

Use Boolean operations such as **Union, Subtract and Intersect** to create a solid form
(Refer to Lesson 3 for step by step instructions)

| **Union** | **Subtract** | **Intersect** |

Shapes can be: **Revolved**   **Shelled**

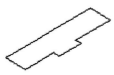  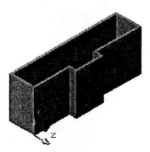

Note: To return to a wireframe display, select the 2D or 3D icon.

*You will learn all of these features and more in the following Lessons.
But first do the following Exercises to get some practice with
Wireframe and Surface Modeling.*

EXERCISE 13A
Create a Wireframe Model

1. Open **My Decimal Set up**

2. Select the **Model tab**

3. **Create the "Wireframe" Model shown below.**
   (Refer to page 13-6 for instructions if necessary)

4. Select **View / Hide**

   *Notice **nothing changed**. Why? Because this is a "wireframe" model.*
   *There are **no surfaces** to "hide" anything.*

5. Save as **EX-13A**

6. **Do not Dimension**

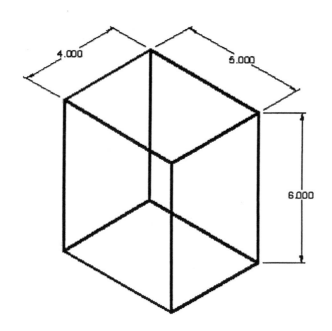

EXERCISE 13B
<u>Create a Surface Model</u>

1. Open **13A**

2. Add Surfaces to all 6 sides.
   Use 3d Orbit to rotate the model to access all sides.

3. Select **View / Hide**

4. Save as **EX-13B**

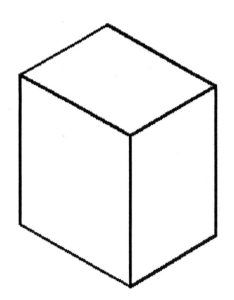

LEARNING OBJECTIVES

After completing this lesson, you will be able to:

1. Construct 6 Solid model Primitives:
    Box, Sphere, Cylinder, Cone, Wedge and Torus

LESSON 14

CONSTRUCTING SOLID PRIMITIVES

AutoCAD has 6 Solid Primitives.
The 6 are, Box, Sphere, Cylinder, Cone, Wedge and Torus.

How to select a solid primitive.

TYPE = TYPE THE NAME
PULLDOWN = DRAW / SOLIDS
TOOLBAR = SOLIDS

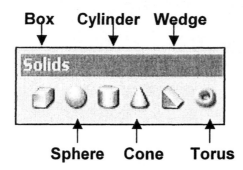

In this lesson you will learn the required steps to construct each of AutoCAD's Solid Primitive objects. Each one requires different input information and some have multiple methods of construction.

3D input direction
When drawing in 3D, and prompted for Length, Width or Height, each input corresponds to an axis direction as follows:

Length = X Axis
Width = Y Axis
Height = Z Axis

For example, if you are prompted for the Length the dimension that you enter will be drawn in the X axis. So keep your eye on the UCS icon in order to draw the objects in the correct orientation.

Consider starting the primitives on the Origin. It is useful to know where the primitive is located so it can be moved or rotated easily.

In Lesson 15 you will learn more about moving the UCS around to fit your construction needs. But relax and let's take it one step at a time.

BOX

There are 4 methods to draw a Solid Box. Which one you will use will depend on what information you know. For example, if you know where the corners of the base are located and the height, then you could use method 1 or 2.

Method 1 (Enter the location for a corner, opposite corner and the height)
a. Select the **SE Isometric** view.
b. Select the **Box** command.
c. Specify corner of box or [CEnter] <0,0,0>: *type coordinates or pick location with cursor. (P1)*
d. Specify corner or [Cube/Length]: *type coordinates for the opposite corner or pick location with the cursor. (P2)*
e. Specify height: *type the height*

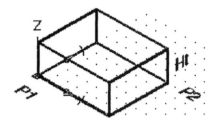

Method 2 (Enter the dimension for L, W, and Ht)
a. Select the **SE Isometric** view.
b. Select the **Box** command.
c. Specify corner of box or [CEnter] <0,0,0>: *type coordinates or pick location with cursor. (P1)*
d. Specify corner or [Cube/**Length**]: *type "L" <enter>*
e. Specify length: *enter the Length (X axis)*
f. Specify width: *enter the Width (Y axis)*
g. Specify height: *enter the Height (Z axis)*

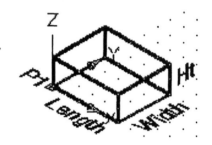

Method 3 (Length, Width & Height have the same dimension)
a. Select the **SE Isometric** view.
b. Select the **Box** command.
c. Specify corner of box or [CEnter] <0,0,0>: *type coordinates or pick location with cursor. (P1)*
d. Specify corner or [**Cube**/Length]: *type "C" <enter>*
e. Specify length: *enter the dimension*

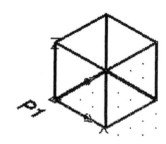

Method 4 (Enter the location for the center, a corner and the height)
a. Select the **SE Isometric** view.
b. Select the **Box** command.
c. Specify corner of box or [CEnter] <0,0,0>: *type "CE"<enter>*
d. Specify center of box <0,0,0>: *type coordinates or pick location with cursor. (P1)*
e. Specify corner or [Cube/Length]: *type coordinates for a corner or pick location with the cursor. (P2)*
f. Specify height: *type the height*

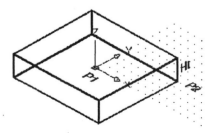

14-3

SPHERE

Sphere creates a spherical solid. You will define center point and then define the size by entering either the radius or the diameter.

a. Select the **SE Isometric** view.
b. Select the **Sphere** command.
c. Specify center of sphere <0,0,0>: *type coordinates or pick location with cursor*
d. Specify radius of sphere or [Diameter]: *enter radius or D*

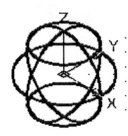

Select **VIEW / HIDE or Type: hi <enter>**

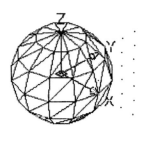

Select **VIEW / SHADE / GOURAUD SHADED**

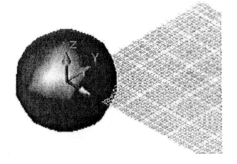

CYLINDER

Cylinder create a cylindrical solid. You will define the center location and then define the size by entering the radius or diameter and the height.

There are 2 methods to specify the height.

Method 1
The default orientation of the cylinder has the base on the UCS plane and the height is in the Z direction. In other words, the diameter is the X or Y axis and the height is the Z axis. When you enter the height the center of the end of the cylinder, extends on the Z axis.
So remember, keep an eye on the UCS icon.

a. Select the **SE Isometric** view.
b. Select the **Cylinder** command.
c. Specify center point for base of cylinder or [Elliptical] <0,0,0>: *type coordinates or pick location with cursor (P1)*
d. Specify radius for base of cylinder or [Diameter]: *enter radius or D*
e. Specify height of cylinder or [Center of other end]: *enter the height (Ht)*

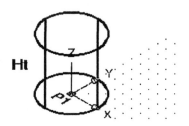

Method 2
The orientation of the Cylinder base depends on the placement of the Center of the other. Define the center of the base and radius then select the "Center of other End" option. Define the "Center of the other end" using coordinates or snapping to an object.

a. Select the **SE Isometric** view.
b. Select the **Cylinder** command.
c. Specify center point for base of cylinder or [Elliptical] <0,0,0>: *type coordinates or pick location with cursor (P1)*
d. Specify radius for base of cylinder or [Diameter]: *enter radius or D*
e. Specify height of cylinder or [Center of other end]: *C <enter>*
f. Specify center of other end of cylinder: *type coordinates or snap to an object. (P2)*
   (This is why starting the object on the Origin is helpful)

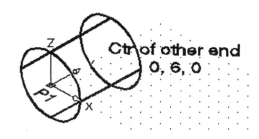

Enter coordinates

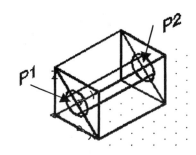

Snap to an object

CONE

Cone creates a Conical solid. There are 2 methods to create a Cone.
You will define the center location and radius or Diameter for the base and then define the height or location for the apex.

Method 1
The default orientation for the base is on the X and Y plane and the height is perpendicular in a Z direction.

a. Select the **SE Isometric** view.
b. Select the **Cone** command.
c. Specify center point for base of cone or [Elliptical] <0,0,0>: *type coordinates or pick location with cursor (P1)*
d. Specify radius for base of cone or [Diameter]: *enter radius or D*
e. Specify height of cone or [Apex]: *enter the apex height*

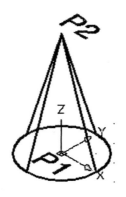

Method 2
The orientation of the Cone depends on the placement of the Apex. Define the center of the base and radius then select the "Apex" option. Define the "Apex" location using coordinates or snapping to an object.

a. Select the **SE Isometric** view.
b. Select the **Cone** command.
c. Specify center point for base of cone or [Elliptical] <0,0,0>: *type coordinates or pick location with cursor (P1)*
d. Specify radius for base of cone or [Diameter]: *enter radius or D*
e. Specify height of cone or [Apex]: *type "A" <enter>*
f. Specify apex point: *type coordinates or snap to an object. (P2)*

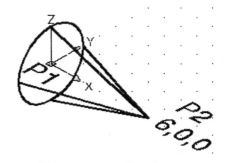
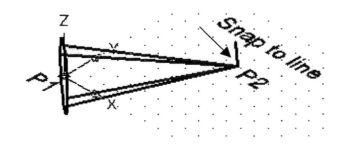

Enter coordinates   Snap to an object

WEDGE

Wedge creates a wedge solid. There are 4 methods to create a Wedge.
The base is always parallel with the current UCS, XY plane, and the slope is always from the Z axis along the X axis.

Method 1 (Define the location for the corners of the base and then the height)
a. Select the **SE Isometric** view.
b. Select the **Wedge** command.
c. Specify corner of box or [CEnter] <0,0,0>: *type coordinates or pick location with cursor. (P1)*
d. Specify corner or [Cube/Length]: *type coordinates for the opposite corner or pick location with the cursor. (P2)*
e. Specify height: *type the height*

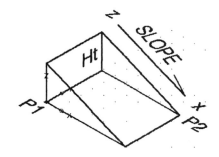

Method 2 (Define the location for each of the three dimensions, L, W, Ht)
a. Select the **SE Isometric** view.
b. Select the **Wedge** command.
c. Specify corner of box or [CEnter] <0,0,0>: *type coordinates or pick location with cursor. (P1)*
d. Specify corner or [Cube/**Length**]: *type "L" <enter>*
e. Specify length: *enter the Length (X axis)*
f. Specify width: *enter the Width (Y axis)*
g. Specify height: *enter the Height (Z axis)*

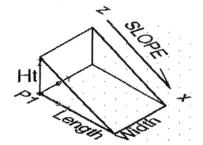

Method 3 (Define Length, Width & Height the same dimension)
a. Select the **SE Isometric** view.
b. Select the **Wedge** command.
c. Specify corner of box or [CEnter] <0,0,0>: *type coordinates or pick location with cursor. (P1)*
d. Specify corner or [**Cube**/Length]: *type "C" <enter>*
e. Specify length: *enter the dimension*

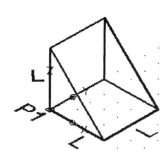

Method 4 (Define the location for the center and the height)
a. Select the **SE Isometric** view.
b. Select the **Wedge** command.
c. Specify corner of box or [**CE**nter] <0,0,0>: *type "CE"<enter>*
d. Specify center of box <0,0,0>: *type coordinates or pick location with cursor. (P1)*
e. Specify corner or [Cube/Length]: *type coordinates for a corner or pick location with the cursor. (P2)*
f. Specify height: *type the height*

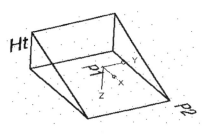

TORUS

Torus can be used to create 3 different solid shapes.

Note: Examples below shown with Hide On.

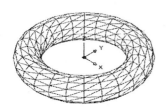

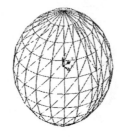

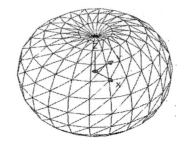

Donut shaped   **Football shaped**   **Self-Intersecting**

The 2 dimensions that are required are the radius
or diameter of the **Torus** and the **Tube**.

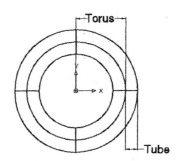

Donut shaped
Note: the <u>Torus radius</u> must be <u>greater than</u> the <u>Tube radius</u>.
a. Select the **SE Isometric** view.
b. Select the **Torus** command.
c. Specify center of torus <0,0,0>: *type coordinates or pick location with cursor. (P1)*
d. Specify radius of torus or [Diameter]: *(this dim. must be <u>greater than</u> the <u>Tube</u> radius)*
e. Specify radius of tube or [Diameter]: *(this dim. must be <u>less than</u> the <u>Torus</u> radius)*

Football shaped
Note: the <u>Torus radius</u> must be <u>negative</u> and the <u>Tube radius positive</u> and <u>greater than</u> the <u>Torus radius</u>.
a. Select the **SE Isometric** view.
b. Select the **Torus** command.
c. Specify center of torus <0,0,0>: *type coordinates or pick location with cursor. (P1)*
d. Specify radius of torus or [Diameter]: *(this dim. must be <u>negative</u>)*
e. Specify radius of tube or [Diameter]: *(this dim. must be <u>positive</u> and <u>greater than the Torus radius</u>)*

Self-Intersecting
Note: the <u>Torus radius</u> must be <u>less than</u> the <u>Tube radius</u>.
a. Select the **SE Isometric** view.
b. Select the **Torus** command.
c. Specify center of torus <0,0,0>: *type coordinates or pick location with cursor. (P1)*
d. Specify radius of torus or [Diameter]: *(this dim. must be <u>less than</u> the <u>Tube</u> radius)*
e. Specify radius of tube or [Diameter]: *(this dim. must be <u>greater than</u> the <u>Torus</u> radius)*

EXERCISE 14A
Create 4 Solid Boxes

1. Open **My Decimal Setup**

2. Select the **Model tab** and **SW Isometric view**

3. **Create 4 solid boxes as shown below.** *You decide which method to use.*
   (Refer to page 14-3 for instructions if necessary)

4. Save as **EX-14A**

5. **Do not Dimension**

BOX A   **BOX B**
L = 14 W = 6 HT = -1   HT = 2

BOX C   **BOX D**
ALL SIDES 2"   CENTER IS 13, 4, .5
                                  HT = 1

REMEMBER   X AXIS = L   Y AXIS = W   Z AXIS = HT

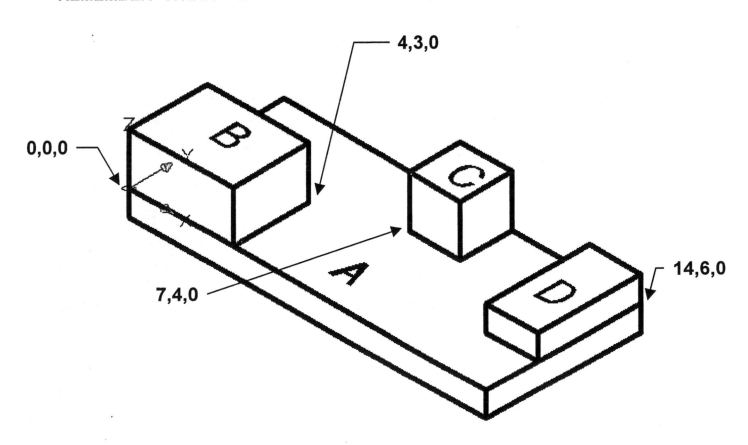

EXERCISE 14B
Create a solid Sphere

1. Open **My Decimal Setup**

2. Select the **Model tab** and **SW Isometric view**

3. **Create the solid Sphere shown below.**
   (Refer to page 14-4 for instructions if necessary)

4. Hide and Gouraud shade

5. Save as **EX-14B**

   Do not Dimension

Center = 0, 0, 0
Radius = 4

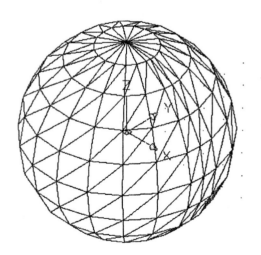

EXERCISE 14C
Create 3 solid Cylinders

1. Open **14A**

2. Add the 3 Cylinders as shown below.
   (Refer to page 14-5 for instructions if necessary)

3. Save as **EX-14C**

   Do not Dimension

CYLINDER E
Radius = 1.25
Ht = 4"

CYLINDER G
6" from end to end
Dia = 1

CYLINDER H
Radius = 1

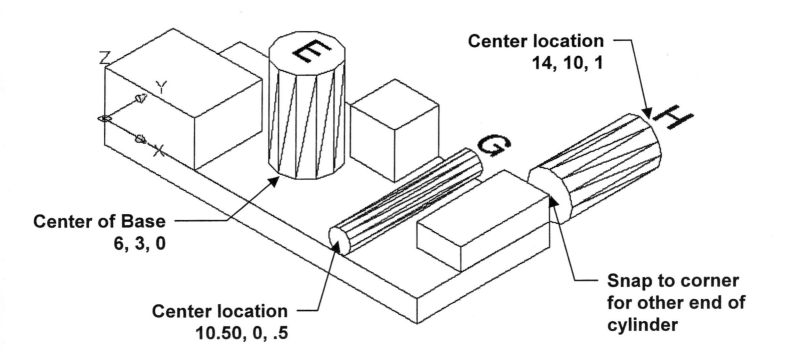

Center location
14, 10, 1

Center of Base
6, 3, 0

Center location
10.50, 0, .5

Snap to corner
for other end of
cylinder

EXERCISE 14D
Create 2 solid Cones

1. Open **14C**

2. Add the 2 Cones as shown below.
   (Refer to page 14-6 for instructions if necessary)

3. Save as **EX-14D**

   Do not Dimension

CONE J	**CONE K**
Ht = 4"	Ht = 2

Note:
Locate the Center location by snapping to the Center of the Cylinder.
Locate the Radius by snapping to the Quadrant of the Cylinder.

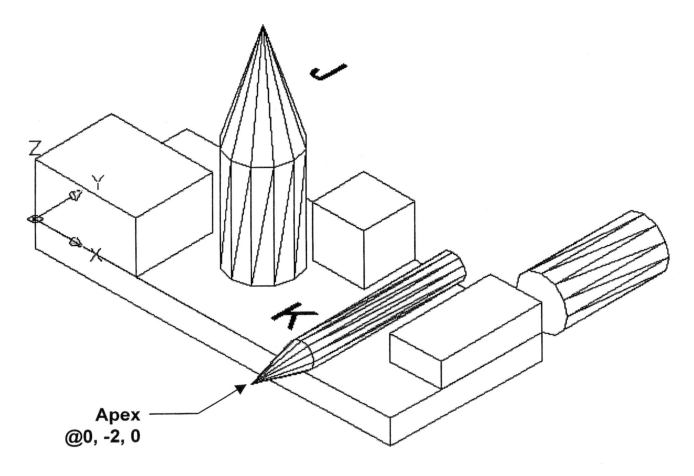

Apex
@0, -2, 0

EXERCISE 14E
Create 3 solid Wedges

1. Open **14D**

2. Add the 3 Wedges as shown below.
   (Refer to page 14-7 for instructions if necessary)

3. Save as **EX-14E**

   Do not Dimension

WEDGE L
Ht = 2"

WEDGE K
L = 1.5
W = 2
Ht = .5

WEDGE H
L, W & H = 2

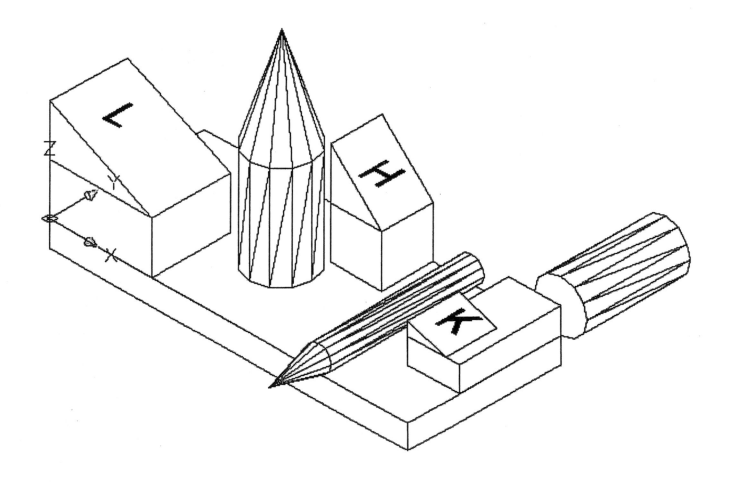

EXERCISE 14F
Create 3 solid Torus'

1. Open **My Decimal Setup**

2. Add the 3 Torus' as shown below.
   (Refer to page 14-8 for instructions if necessary)

3. Save as **EX-14F**

   Do not Dimension

DONUT	**FOOTBALL**	**SELF-INTERSECTING**
Center = 4, 3, 0	Center = 8.5, 7, 0	Center = 10.5, 3, 0
Torus Rad = 3	Torus Rad = -3	Torus Rad = 2
Tube = 1	Tube = 5	Tube = 4

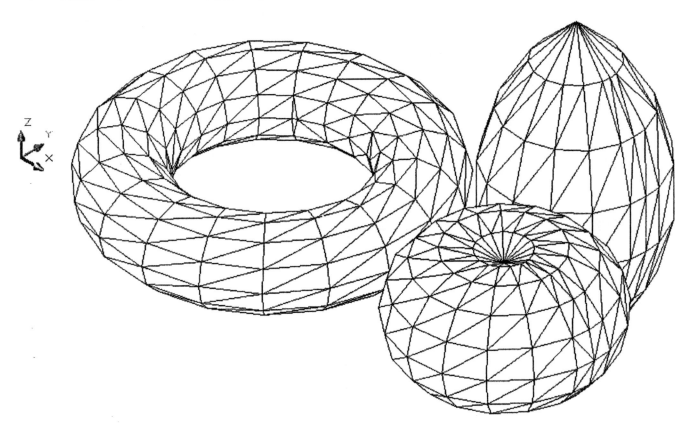

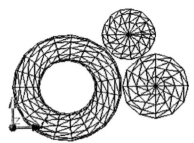

Take a look a the Top View to see if you have them positioned correctly. Select View / 3D Views / Top.

LEARNING OBJECTIVES

After completing this lesson, you will be able to:

1. Move the UCS to aid in the construction of the model.
2. Split the screen into 2 viewports
3. Construct model easier using Plan view.
4. Understand and use Boolean operations:
     Union, Subtract and Intersect

LESSON 15

UNDERSTANDING THE UCS

In this Lesson you are going to learn how to manipulate the <u>UCS Origin</u> to make constructing 3D models easy and accurate.

World Coordinate System vs. User Coordinate System
All objects in a drawing are defined with XYZ coordinates measured from the 0,0,0 Origin. This coordinate system is **fixed** and is referred to as the **World Coordinate System (WCS)**. When you first launch AutoCAD the WCS icon is in the lower left corner of the screen. (An easy way to return to WCS is: Type UCS <enter> <enter>)

On the other hand, the **User Coordinate Sytem (UCS)** can move it's Origin to any location. (This is a procedure, with which you are familiar, referred to as "moving the Origin")

Why move the UCS Origin?
Objects that are drawn will always be parallel to the XY plane. (Unless you type a Z coordinate) So it is necessary to define which plane you want to work on. You define the plane by moving the UCS origin to a surface.
Let me give you a few examples and maybe it will become clearer.

Example 1:
This wedge was drawn using L, W, H. When you entered the L, W and Ht how did it know to draw its base parallel to the XY plane? Because all objects are drawn parallel to the XY plane unless you enter a Z coordinate. Length is always in the X axis, Width is always in the Y axis and Height is in the Z axis. Notice the UCS icon displays the plane orientation.

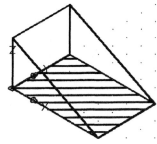

Example 2:
Now I have drawn a cylinder parallel to the sloped surface. How did I do that? Notice the UCS icon. I moved it to the surface I wanted to draw on. (Remember all objects are drawn parallel to the XY plane, unless you enter a Z coordinate.) The base of the cylinder is automatically placed on the XY surface I defined and the height is always drawn in the Z axis.

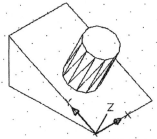

How did I move the UCS icon to the orientation shown in Example 2?
I will discuss that next.

MOVING the UCS ICON

There are many options available to manipulate the UCS icon. In this Lesson I will only be discussing a few. When you get more experienced with AutoCAD's 3D world you may wish to investigate other options.

MOVING THE UCS ICON

Method 1 – Move UCS
You should be very familiar with this method. You simply select **Tools / Move UCS** and select the new location by typing coordinates, from the existing Origin, or by placing it with the cursor.

Method 2 - 3 Point
This method allows you to select in which direction you want the X and Y axes pointing. This is primarily used to attach the UCS to an existing surface. (Such as the slope in Example 2 on the previous page)

1. Select the 3 point command using one of the following:

   > **TYPE = UCS / New / 3point**
   > **PULLDOWN = Tools / New UCS / 3 point**
   > **TOOLBAR = UCS**

2. Specify new origin point <0,0,0>: *Locate where you want the new 0,0,0 (type coordinates or place with the cursor)*

3. Specify point on positive portion of X-axis <1.000,0.000,0.000>: *define which direction is positive X axis by typing coordinates or with the cursor.*

4. Specify point on positive-Y portion of the UCS XY plane <0.000,1.000,0.000>: *define which direction is positive Y axis by typing coordinates or with the cursor.*

<u>**EXAMPLES**</u>

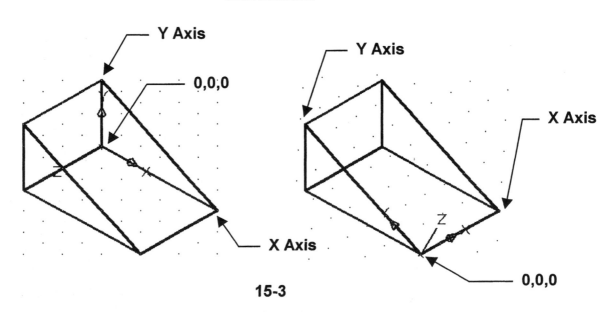

ROTATING the UCS ICON

Rotating the UCS is another option to manipulate the UCS icon. You may rotate around any one of the 3 axes. You may use the cursor to define the rotation angle or type the rotation angle.

1. Select the UCS Rotate command using one of the following:

>   **TYPE = UCS / New / select X, Y or Z**
>   **PULLDOWN = Tools / New UCS / select X, Y or Z**
>   **TOOLBAR = UCS**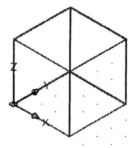

2. Select which Axis to rotate "**Around**"
   *(For Example: If you select **X**, you are rotating the Y and Z axes around the X axis)*

3. Specify rotation angle about X axis <90>: ***type the rotation angle or use cursor***

Understanding the rotation angle
The easiest way to determine the rotation angle is to think of yourself standing at the point of the axis arrow about to be pierced in the chest.
For example, if you are rotating around the X axis, put the point of the X axis arrow against your chest and then grab the Z axis with one hand and the Y axis with your other hand. Now rotate your hands Clockwise or Counterclockwise like you are holding a wheel vertically. The X axis arrow, corkscrewing into your chest, is the Axis you are rotating around. CW is negative and CCW is positive just like the Polar clock you learned about in the Beginning workbook.

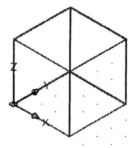

Original UCS Position 1

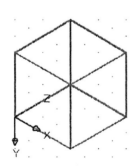

X axis Rotation -90

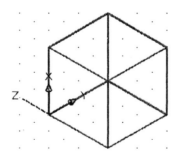

Y axis Rotation -90

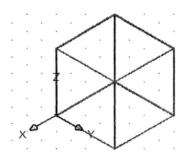

Z axis Rotation -90

Note: To return the UCS to its default location and rotation type UCS <enter> <enter>.
If its stubborn select the SE Isometric View and then type UCS <enter> <enter>

NEW DIRECTION FOR Z AXIS

Remember that the Height dimension is always the positive Z axis. Well sometimes the positive Z axis is oriented to suit your height need. The **Zaxis** option allows you to change the positive direction of the Z axis and the X and Y axis will automatically change also.

1. Select the **Zaxis** command using one of the following:

> **TYPE = UCS / New / Zaxis**
> **PULLDOWN = UCS / NEW / Z axis Vector**
> **TOOLBAR = UCS** ⌊Z⌉

2. Specify new origin point <0,0,0>: *type the coordinates or place the cursor*

3. Specify point on positive portion of Z-axis <0.000,0.000,1.000>: *type the coordinates or place the cursor*

EXAMPLES
Notice all 3 examples use the same information to draw the Cylinder. The only difference is the "drawing plane".

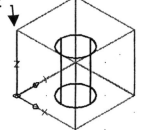

The Original
The XY drawing plane is on the <u>inside bottom</u>.
I draw a Cylinder using the following:
Base center location: **2, 2, 0**
Radius : **1**
Ht = **3**

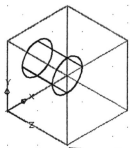

Positive Zaxis direction has been changed
The XY drawing plane is on the <u>inside of the left side</u>.
I draw a Cylinder using the following:
Base center location: **2, 2, 0 (same as above)**
Radius : **1 (same as above**
Ht = **3 (same as above**

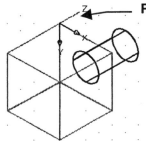

Positive Zaxis direction has been changed again
The XY drawing plane is on the <u>outside of the Back</u>.
I draw a Cylinder using the following:
Base center location: **2, 2, 0 (same as above**
Radius : **1 (same as above**
Ht = **3 (same as above**

DRAWING WITH TWO VIEWPORTS

Sometimes it is easier to work on the model if you split Model space into two viewports. An isometric view can be displayed in one and the plan view in the other. This makes it easy to move the UCS around in the isometric view and draw on the XY plane in the plan view.

Setting the UCS ICON system variable

The **UCSVP** system variable controls the UCS icon display configuration.

0 The UCS icon, in each viewport, is relative to each other. If you change one, they all change.

1 The UCS icon, in each viewport is independent. Each viewport may have it's own UCS orientation. (This is the default setting)

I find it less confusing, for students new to 3D, to use the "0" setting.
If you change one, you change them all.
*So I want you to change the **UCSVP** variable to "0".*

1. *Type: ucsvp<enter>*
2. *Type: 0 <enter>*

You should change this system variable on all of your master setup drawings then resave the drawing. Then you do not have to worry about this until you decide you want to change it to "1".
(Do this now so you don't forget, but I will remind you for awhile)

HOW TO SPLIT THE SCREEN INTO 2 VIEWPORTS
1. Check the **ucsvp** setting. (I will continue to remind you)
2. Select **VIEW / VIEWPORTS / 2 VIEWPORTS**
3. Enter a configuration option [Horizontal/Vertical] <Vertical>: *press <enter>*
4. Activate the right hand viewport. (Click in it)
5. Select **SE Isometric** view. (Notice the UCS icon location)
6. Activate the left hand viewport (Click in it)
7. Select the **VIEW / 3D VIEWS / TOP**

Now think about this. With the left viewport still active, select some of the other views such as, Bottom or Side. Watch the UCS icon move around on the isometric view to match your selection in the left viewport.
If you move the UCS icon in the right hand viewport (isometric view) it will also reflect the change within the left hand viewport.
They are relative to each other.

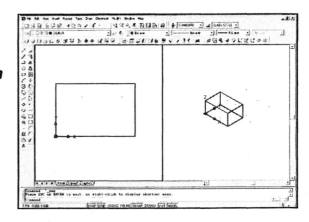

To return to 1 viewport, select **VIEW / VIEWPORTS / 1 VIEWPORT.**

On the next page I will discuss how this will assist you when constructing a 3D model.

PLAN VIEW

Now that you have learned to split the screen into 2 viewports, let's see how you can use this to assist you with constructing a 3D model.

Remember, you always draw on the XY plane.

HOW TO USE THE "PLAN" VIEW
1. Set **UCSVP** to "**0**" (refer to page 15-6)
2. Split the screen into 2 viewports (refer to page 15-6)
3. Activate the right hand viewport. (isometric view)
4. Move the UCS icon to the plane that you wish to draw on.
5. Activate the left hand viewport
6. Type: **PLAN <enter>**
7. Enter an option [Current ucs/Ucs/World] <Current>: **<enter>**

The left hand viewport will display the current XY plane.
Now you can draw on the XY plane as if you were drawing a 2D drawing.
(Note: Each viewport may be zoomed and panned independently.)

EXAMPLES

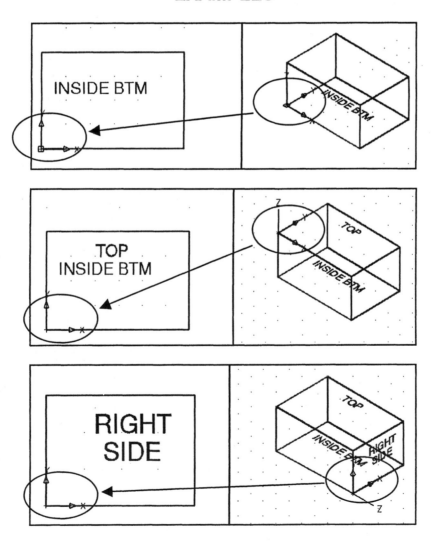

BOOLEAN OPERATIONS

Solid objects can be combined using Boolean operations to create *composite solids*. AutoCAD's Boolean operations are **Union, Subtract** and **Intersect.**

UNION

The **Union** command creates one solid from 2 or more solid objects.

1. Select the **Union** command using one of the following:
   TYPE = UNI or UNION
   PULLDOWN = MODIFY / SOLIDS EDITING / UNION
   TOOLBAR = SOLIDS EDITING

2. Select objects: *select the solids to be combined*

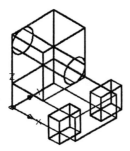 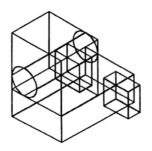 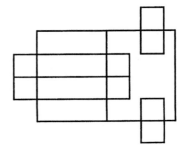

BEFORE UNION

SE Isometric   **SW Isometric**   **Top View**

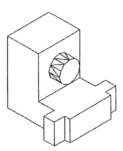 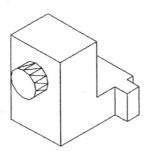 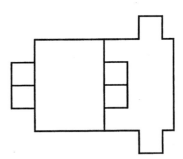

AFTER UNION
(Hide has been used on the isometric views)

SUBSTRACT

The **Subtract** command subtracts one solid from another solid.

1. Select the **Subtract** command using one of the following:
   TYPE = SU or SUBTRACT
   PULLDOWN = MODIFY / SOLIDS EDITING / SUBTRACT
   TOOLBAR = SOLIDS EDITING

2. Select solids and regions to subtract from…
   Select objects: *select the solid object to subtract <u>from</u>*
   Select objects: *<enter>*

3. Select solids and regions to subtract…
   Select objects: *select the solid object to subtract*
   Select objects: *<enter>*

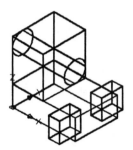
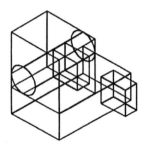
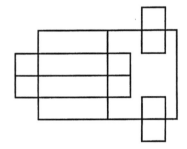

BEFORE SUBTRACT

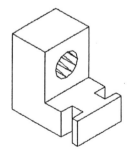
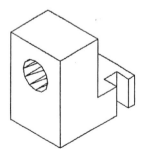
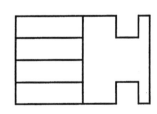

AFTER SUBTRACT
(Hide has been used on the isometric views)

INTERSECTION

If solid objects intersect, they <u>share a space</u>. This shared space is called an Intersection. The **Intersection** command allows you to create a solid from this shared space.

1. Select the **Intersection** command using one of the following:
   TYPE = IN or INTERSECT
   PULLDOWN = MODIFY / SOLIDS EDITING / INTERSECT
   TOOLBAR = SOLIDS EDITING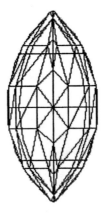

2. Select objects: *select the solid objects that form the intersection*
   Select objects: *<enter>*

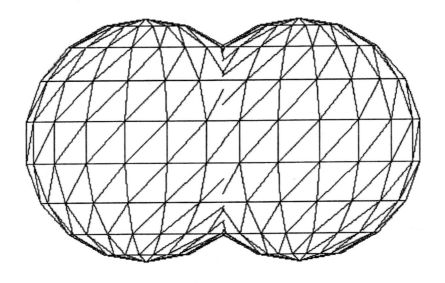

BEFORE INTERSECTION

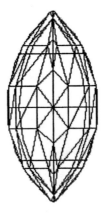

AFTER INTERSECTION

EXERCISE 15A
Subtract

1. Open **My Decimal Setup**

2. Select the **Model tab** and **SW Isometric view**

3. Check the **UCSVP** setting. It should be "0"
   (very important if you are using multiple viewports)

4. Draw the solid object shown below.

5. Save as **EX-15A**

6. **Do not Dimension**

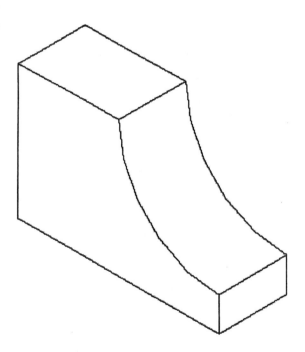

Step by step instructions and dimensions are shown on the next page.

Note: There are many different methods to create the object above.
The steps on the next page are designed to make you think about UCS
positioning and rotating and the positive and negative inputs.
Can you think of other methods?

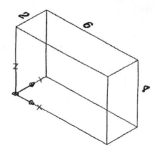

Step 1.
Create a solid box.

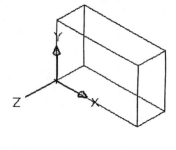

Step 2.
Rotate the UCS icon

Cylinder Center Point

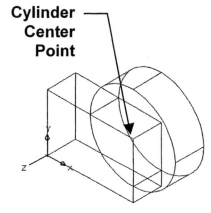

Step 3.
Draw a 3" Radius Cylinder
(Note: Ht = negative 2)

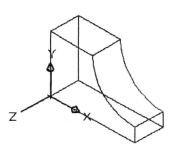

Step 4.
Subtract the Cylinder from the box

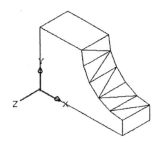

Step 5.
Hide

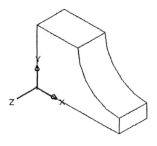

Step 6.
Set Dispsilh to 1 (ref. page 13-5)
Hide again

EXERCISE 15B
Union and Subtract

1. Open **15A**

2. Check the **UCSVP** setting. It should be "0"
   (very important if you are using multiple viewports)

3. Add to the existing model to form the solid object below.

4. Save as **EX-15B**

5. **Do not Dimension**

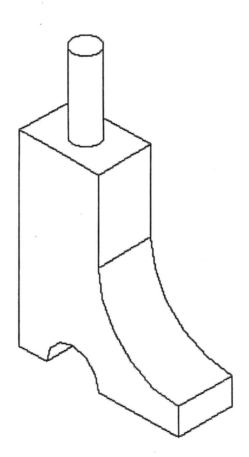

A few construction hints and dimensions are shown on the next page.

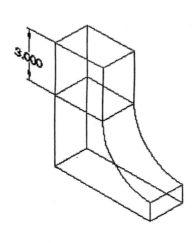

Step 1.
Add a solid box.

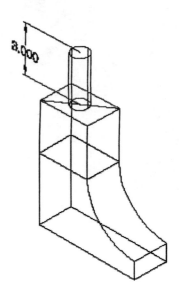

Step 2.
Add a cylinder

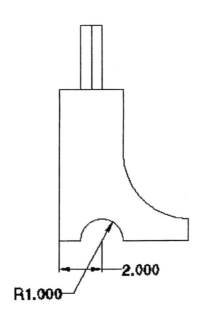

Step 3.
Subtract a cylinder
Union and hide

Note: Are you having problems with dispsilh? Does it work sometimes and other times not? That is normal. Try selecting the 3D wireframe icon and then the 2D wireframe button. Then try hide again.

EXERCISE 15C
Moving the UCS

1. Open **My Decimal Setup**

2. Check the **UCSVP** setting. It should be "0"
   (very important if you are using multiple viewports)

3. Draw the 6" solid cube shown below

4. Add the text (Height = .50) on the outside of each side.

5. Save as **EX-15C**

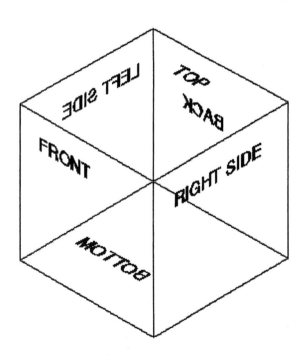

**This is a perfect exercise for using 2 viewports and the "Plan" view.
Use Isometric views and 3D Orbit to spin the cube around.
Use "3point" to move the Origin to reposition the XY plane**

EXERCISE 15D
Assembling 3D solids

1. Open **My Feet-Inches Setup**

2. Select the **Model** tab

3. Check the **UCSVP** setting. It should be "0"
   (very important if you are using multiple viewports)

4. Draw the table shown below

5. Save as **EX-15D**

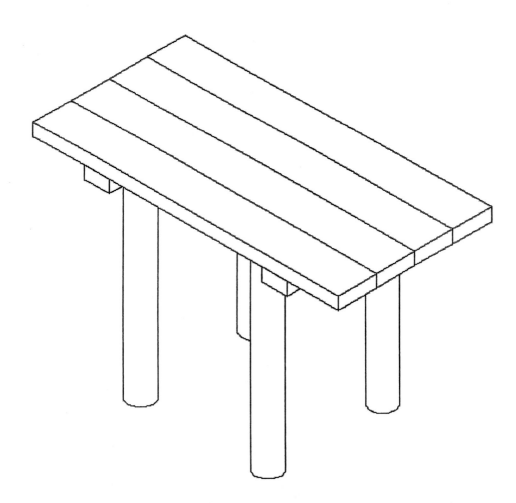

Dimensions and construction hints on the next page.
Have some fun with this.
If you have time, add some benches.

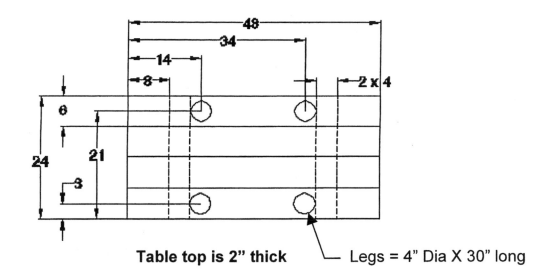

Table top is 2" thick   Legs = 4" Dia X 30" long

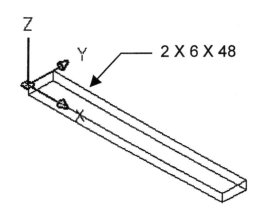

2 X 6 X 48

Start with a plank.

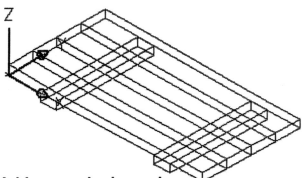

**Add more planks and
the 2 x 4's (Notice the UCS position)**

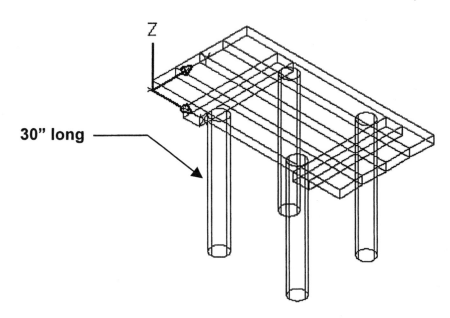

30" long

**Add the <u>30" long</u> legs
Remember, to make the legs go in the negative Z axis**

NOTES:

LEARNING OBJECTIVES

After completing this lesson, you will be able to:

1. Extrude solid surfaces.
2. Create a Region

LESSON 16

EXTRUDE

AutoCAD's primitives are helpful in constructing 3D solids quickly. But what if you want to create a solid object that has a more complex shape, such as the one shown here.

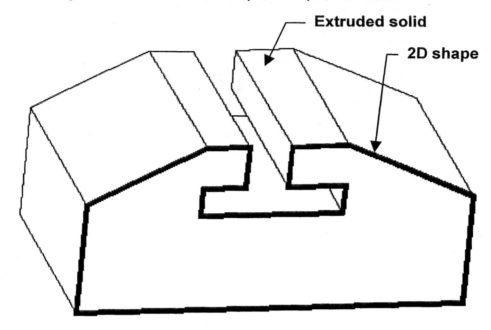

Extruded solid

2D shape

To create the solid shown above you first need to draw a closed 2D shape and then **EXTRUDE** it.

The **EXTRUDE** command allows you to take a 2D shape and extrudes it into a solid. It extrudes <u>along the Z-Axis.</u> Extrusions can be created along a straight line or along a curved path. The extrusion may also have a taper.

Only **closed** 2D shapes such as Circle, Polygon, Rectangle, Ellipse, Donuts, <u>closed</u> Polylines and Regions can be extruded. (Splines can also be extruded but we are not discussing those in this workbook)

There are 4 methods for extruding a 2D shape:
1. Perpendicular to the 2D shape with straight sides.
2. Perpendicular to the 2D shape with tapered side.
3. Along a Path
4. Extrude a Region.

HOW TO SELECT THE EXTRUDE COMMAND

> **TYPE = EXT or EXTRUDE**
> **PULLDOWN = DRAW / SOLIDS / EXTRUDE**
> **TOOLBAR = SOLIDS**

EXTRUDE - <u>PERPENDICULAR WITH STRAIGHT SIDES</u>

1. Draw a closed 2D shape on the XY plane.

2. Select the **EXTRUDE** command using one of the methods shown on page 16-2.

3. Select the objects: *select the objects to extrude <enter>*

4. Select the objects: *<enter>*

5. Specify height of extrusion or [Path]: *type the height along the Z axis <enter>*

    (Note: a positive value extrudes above the XY plane and negative extrudes below the XY plane.)

6. Specify angle of taper for extrusion <0>: *<enter>*

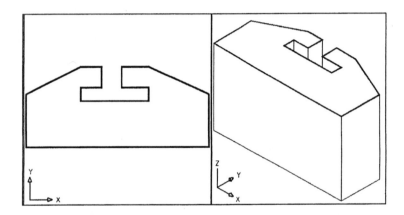

Note the orientation of the extrusion in the left viewport.
This may not be the orientation you wanted.
Remember the object will extrude on the Z axis.
You may find it necessary to rotate the UCS before creating the drawing.

This is a great time to use the **PLAN** method described on page 15-7.

Before drawing the 2D shape do the following:
1. Set the UCSVP setting to "0"
2. Split the screen into 2 viewports
3. Set the right hand viewport to **SE Isometric** view.
4. Rotate the UCS if necessary
5. Set the left hand viewport to **PLAN** view
6. Draw the 2D shape in the left hand viewport.
7. **Extrude**

EXTRUDE - <u>PERPENDICULAR WITH TAPERED SIDES</u>

1. Draw a closed 2D shape on the XY plane.
2. Select the **EXTRUDE** command
3. Select the objects: ***select the objects to extrude <enter>***
4. Select the objects: ***<enter>***
5. Specify height of extrusion or [Path]: ***type the height along the Z axis <enter>***
   (Note: a positive value extrudes above the XY plane and negative extrudes below the XY plane.)
6. Specify angle of taper for extrusion <0>: ***enter taper angle <enter>***

How to control the taper direction
If you enter a positive angle the resulting extruded solid will taper inwards.
If you enter a negative angle the resulting extruded solid will taper outwards.

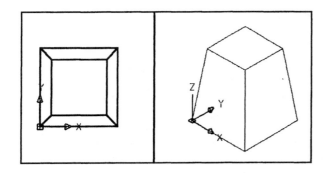   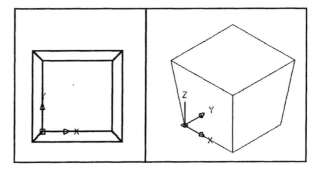

Positive angle   **Negative angle**

EXTRUDE - <u>ALONG A PATH</u>

Extruding a 2D shape along a path is not difficult but there are some definite rules.
a. The path should be drawn perpendicular to the 2D shape.
b. You may use the following objects to draw a path; Line, Arc or 2D polyline
c. The path must have a beginning and an end, so you cannot use a Circle.

1. Draw a closed 2D shape on the XY plane.
2. Draw the path perpendicular to the shape
   (rotate the UCS 90 degrees)
3. Select the **EXTRUDE** command
4. Select the objects: ***select the objects to extrude <enter>***
5. Select the objects: ***<enter>***
6. Specify height of extrusion or [Path]: ***type P <enter>***
7. Select extrusion path or [Taper angle]:***select the path (object)***

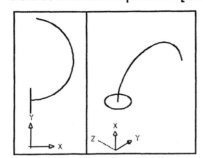   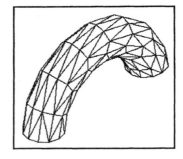

REGION

A **REGION** is a solid with no thickness. Think of a piece of paper. Thin but it is a solid. You can use all the Boolean operations, Union, Subtract and Intersect, on the region <u>and</u> you can <u>extrude</u> it.

Example:

1. I have a 2D drawing of a flat plate with circles on it. I would like to extrude this plate and I want the circles to be actual holes.

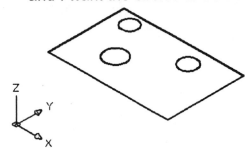

2. I use the **REGION** command to transform these objects into <u>solids</u>.

   a. Select the **REGION** command using one of the following:
      TYPE = REG or REGION
      PULLDOWN = DRAW / REGION
      TOOLBAR = DRAW

   b. Select objects: *select the objects <enter>*
      Select objects: *<enter>*
      4 loops extracted.
      4 Regions created.

3. Now that they are solids I can subtract the circles from the rectangle, so they are actually "holes".

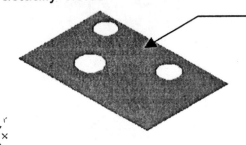

I added shading so you could see that the circles are now holes and have been subtracted from the rectangle.

4. I Extrude the region.

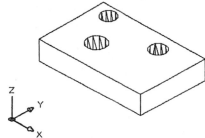

Polyline's "JOIN" option

You learned how to edit Polylines in Lesson 24 of the Exercise Workbook for ***Beginning*** *AutoCAD. But you may not remember Polyline's "JOIN" option. This option is very useful when attempting to extrude a 2D shape, so I would like to discuss it again.*

The **JOIN** option allows you to join multiple polyline segments into one polyline. The JOIN option also will transform a **LINE** into a Polyline. Remember, you can only extrude a **closed** object. *4 LINES, with touching endpoints, is not a closed object.* They are 4 individual LINE segments. The LINE segments must be transformed into polylines and JOINed to create a closed object.

HOW TO USE THE JOIN COMMAND

1. Draw 6 touching lines approximately as shown. Each endpoint much touch.

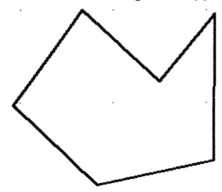

2. Select the Polyedit command using one of the following:
    TYPE = PEDIT
    PULLDOWN = MODIFY / OBJECT / POLYLINE
    TOOLBAR = MODIFY II

3. Command: _pedit Select polyline or [Multiple]: ***select ONE polyline (click on it)***
    Object selected is not a polyline (***if the object is not a polyline this appears***)
    Do you want to turn it into one? <Y> ***Y <enter>***

4. Enter an option [Close/Join/Width/Edit vertex/Fit/Spline/Decurve/Ltype gen/Undo]:***J <enter>***

5. Select objects: ***select first corner of window***

6. Specify opposite corner: ***select opposite corner of window***

7. Select objects: **<enter>**
    5 segments added to polyline

8. Enter an option [Open/Join/Width/Edit vertex/Fit/Spline/Decurve/Ltype gen/Undo]: ***<enter>***

Now you have transformed 6 LINES into a closed polyline. You are now able to extrude this shape.

EXERCISE 16A
Extrude

1. Open **My Decimal Setup**
2. Select the **Model tab**
3. Draw the 2D drawing approximately as shown below to form a closed Polyline. (Fig. 1)
4. Extrude (4") to form the solid object. (Fig. 2)
5. Select the **SE Isometric** view and hide.
6. Save as **EX-16A**

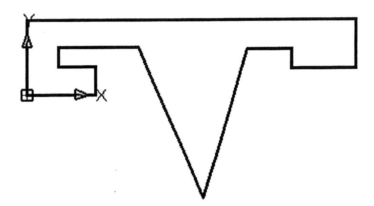

Figure 1

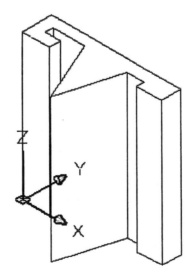

Figure 2

EXERCISE 16B
Extrude with Taper

1. Open **My Decimal Setup**
2. Select the **Model tab**
3. Draw a 6 sided Inscribed Polygon with a radius of 2 (Fig. 1)
4. Extrude to form the solid object. Ht = 4 Taper = 6 (Fig. 2)
5. Add the letter & Number shapes (Use polyline, do not use text).
   Extrude each. Ht = 1 (Fig. 3)
   Don't forget to use subtract on the number shape "4" and "6".

   Refer to the next page for construction suggestions

6. Select the **SE Isometric** view and hide.
7. Save as **EX-16B**

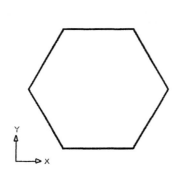

Figure 1

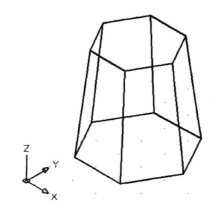

Figure 2

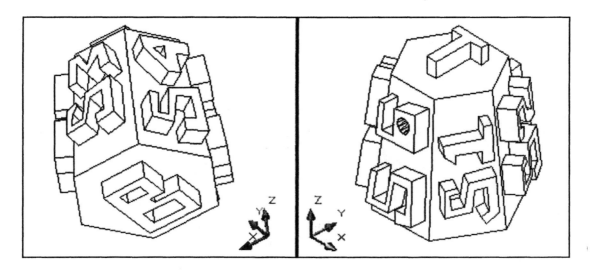

Figure 3

16-8

Construction suggestion:

a. Move the UCS, using 3pt, to select the correct surface to draw on.

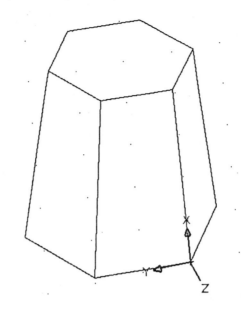

b. View the **current Plan view** to draw the text.
Check the **UCSVP** setting. What should it be set to? 0 or 1?
Type **Plan <enter> <enter>**

c. Extrude the polyline letter shapes.

d. Select **SE Isometric view** and then use **3D Orbit** to rotate around to the next surface.
Use the HIDE command if the objects start visually overlapping.

EXERCISE 16C
Extrude along a Path

1. Open **My Decimal Setup**
2. Select the **Model tab**
3. Draw a 6 sided Inscribed Polygon with a radius of 2 (Fig. 1)
4. Draw a polyline path perpendicular to the Polygon approximately as shown (Fig. 2)
5. Extrude the Polygon along the path (Fig. 3)
6. Select the **SE Isometric** view and hide.
7. Save as **EX-16C**

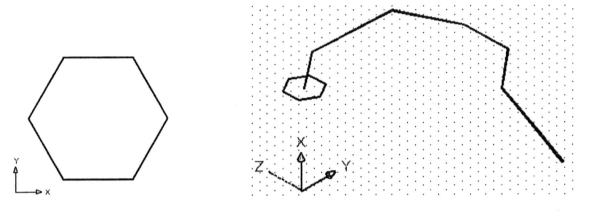

Figure 1

Figure 2

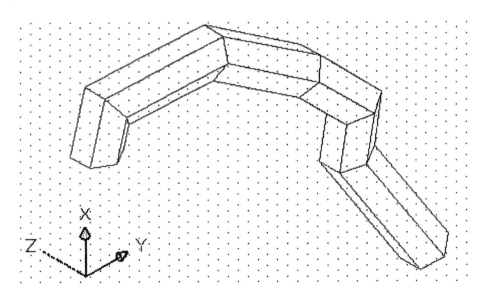

Figure 3

EXERCISE 16D
Extrude a Region

1. Open **My Decimal Setup**
2. Select the **Model tab**
3. Draw the 2D drawing shown below. (Fig. 1)
   Note: Orientation must be correct so you may have to rotate the UCS.
4. Extrude (4") to form the solid object. (Fig. 2)
   Note: Use Region and Subtract
5. Select the **SE Isometric** View and **Hide**.
6. Save as **EX-16D**

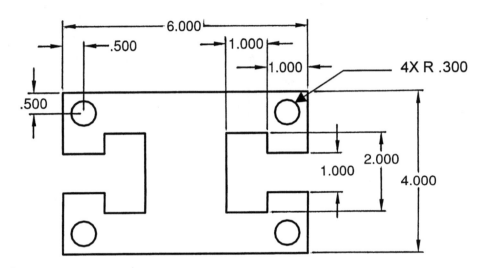

Figure 1

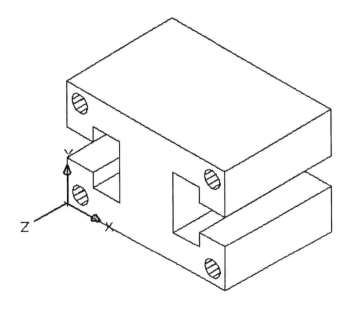

Figure 2

EXERCISE 16E
Join and Extrude a Region

1. Open **EX-5E**
2. Select the **Model tab**
3. **Erase**: Dimensions, Furniture, Symbols, Text and Windows
4. If you used the Multiline command to create this floorplan, **EXPLODE** the floorplan. *(The Extrude command will not work with Multilines)*
5. Select the **SE Isometric** View.
6. Use the "**JOIN**" option (refer to page 16-6) to transform the lines into closed polylines <u>or</u> use the "**REGION**" command (refer to page 16-5) to create 4 regions.

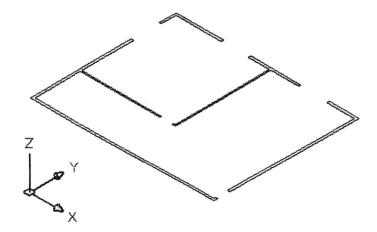

7. EXTRUDE the walls to a height of 8 feet.
8. Save as **EX-16E**

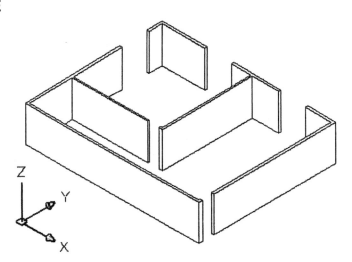

To add walls or windows you can use the Union or Subtract.
Have some fun with this one. Try adding the furniture.

LEARNING OBJECTIVES

After completing this lesson, you will be able to:

1. Understand 3D Operations.
   Mirror 3D, Rotate 3D, Align and 3D Array

LESSON 17

3D OPERATIONS

3D Operations allow you to **Mirror, Rotate, Align** and **Array** a solid. The methods are almost identical to the 2D commands with the exception of defining the plane. You do not have to move the UCS.

Note: You may use the equivalent 2D commands but they will only work in the XY plane. It may be necessary to move the UCS.

THE FOLLOWING ARE EXPLANATIONS FOR THE 3D COMMANDS.

MIRROR 3D
You must define the mirror plane.

1. Select the **MIRROR 3D** command using one of the following:

    TYPE = MIRROR3D
    PULLDOWN = MODIFY / 3D OPERATIONS / MIRROR 3D
    TOOLBAR = NONE

2. Select objects: ***select the solid to be mirrored***

3. Select objects: ***<enter>***

4. Specify first point of mirror plane (3 points) or
    [Object/Last/Zaxis/View/XY/YZ/ZX/3points] <3points>: ***<enter>***

5. Specify first point on mirror plane: ***select the vertex of the plane (P1)***

6. Specify second point on mirror plane: ***select the positive X direction (P2)***

7. Specify third point on mirror plane: ***select the positive Y direction (P3)***

8. Delete source objects? [Yes/No] <N>:***<enter>***

EXAMPLE:

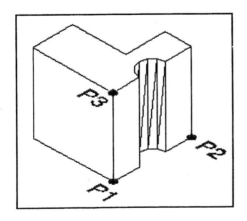 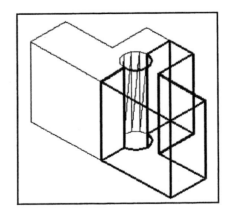

ROTATE 3D

You must pick 2 points to define the axis of rotation and the rotation angle. To determine the rotation angle you must look down the axis from the second point.

If you use my stabbing arrow analogy, you will put the positive end of the second point selected, in your chest and rotate the other two axes. Positive input is counterclockwise and Negative input is clockwise.

1. Select the **ROTATE 3D** command using one of the following:

   TYPE = ROTATE3D
   PULLDOWN = MODIFY / 3D OPERATIONS / ROTATE 3D
   TOOLBAR = NONE

2. Current positive angle: ANGDIR=counterclockwise ANGBASE=0
   Select objects: *select the solid to be rotated*

3. Select objects: *<enter>*

4. Specify first point on axis or define axis by
   [Object/Last/View/Xaxis/Yaxis/Zaxis/2points]: *select first point on axis (P1)*

5. Specify second point on axis: *select second point on axis (P2)*

6. Specify rotation angle or [Reference]: *type rotation angle*

EXAMPLE: Rotated 180 degrees

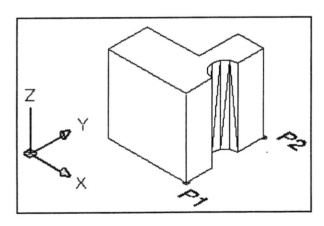
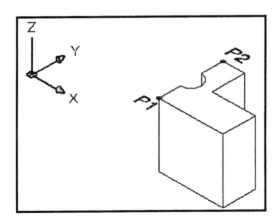

ALIGN

The **ALIGN** command allows you to MOVE and ROTATE an object from its existing location to a new location.

You define the **SOURCE** points (existing location) and the **DESTINATION** points (new location for the source points).

1. Select the **ALIGN** command using one of the following:

    TYPE = AL or **ALIGN**
    PULLDOWN = MODIFY / 3D OPERATIONS / ALIGN
    TOOLBAR = NONE

2. Select objects: *select the object to be aligned*

3. Select objects: *<enter>*

4. Specify first source point: *select the first source (existing point) S1*

5. Specify first destination point: *select the first destination (new location point) D1*

6. Specify second source point: *select the second source (existing point) S2*

7. Specify second destination point: *select the second destination (new location point) D2*

8. Specify third source point or <continue>: *select the third source (existing point) S3*

9. Specify third destination point: *select the third destination (new location point) D3*

EXAMPLE:

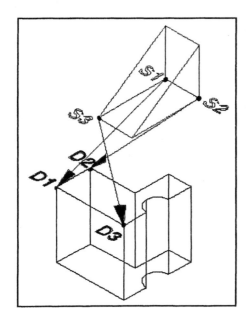
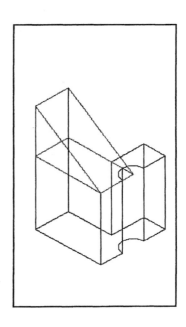

17-4

3DARRAY

RECTANGULAR
You define the:

        ROWS = number of copies needed in the **Y** direction
        COLUMNS = number of copies needed in the **X** direction
        LEVELS = number of copies needed in the **Z** direction
        DISTANCE BETWEEN ROWS, COLUMNS and LEVELS

1. Select the **3DARRAY** command using one of the following:

   TYPE = 3A or 3DARRAY
   PULLDOWN = MODIFY / 3D OPERATIONS / 3DARRAY
   TOOLBAR = NONE

2. Select objects: *select object to be arrayed*

3. Select objects:*<enter>*

4. Enter the type of array [Rectangular/Polar] <R>: *R <enter>*

5. Enter the number of rows (---) <1>: *type number of rows (Ydirection) <enter>*

6. Enter the number of columns (||||) <1>: *type number of columns (X direction) <enter>*

7. Enter the number of levels (...) <1>: *type number of levels (Z direction) <enter>*

8. Specify the distance between rows (---): *type distance between rows<enter>*

9. Specify the distance between columns (||||): *type distance between columns<enter>*

10. Specify the distance between levels (...): *type distance between levels <enter>*

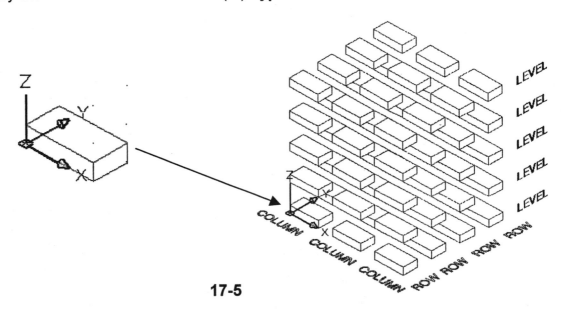

POLAR

The object is arrayed around an <u>entire axis</u> rather than just a point.

You define the:
- **NUMBER OF COPIES**
- **ANGLE TO FILL**
- **ROTATION DIRECTION**
- **ROTATION AXIS ENDPOINTS**

1. Select the **3DARRAY** command using one of the following:

   TYPE = 3A or 3DARRAY
   PULLDOWN = MODIFY / 3D OPERATIONS / 3DARRAY
   TOOLBAR = NONE

2. Select objects: *select object to be arrayed*

3. Select objects: *<enter>*

4. Enter the type of array [Rectangular/Polar] <R>: *P <enter>*

5. Enter the number of items in the array: *type the number of copies (include original)*

6. Specify the angle to fill (+=ccw, -=cw) <360>: *enter angle for copies to fill*

7. Rotate arrayed objects? [Yes/No] <Y>: *copies rotated, yes or no?*

8. Specify center point of array: <Osnap on>*snap to first endpoint of axis (P1)*

9. Specify second point on axis of rotation:*snap to second endpoint of axis (P2)*

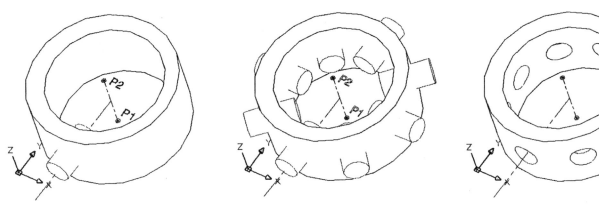

| ORIGINAL | POLAR ARRAY | SUBTRACT |

EXERCISE 17A

MIRROR 3D

1. Open **My Decimal Setup**
2. Select the **Model tab**
3. Draw the solid object below.
4. Dimensions and construction hints on page 17-8. (Consider using MIRROR 3D)
5. Save as **EX-17A**

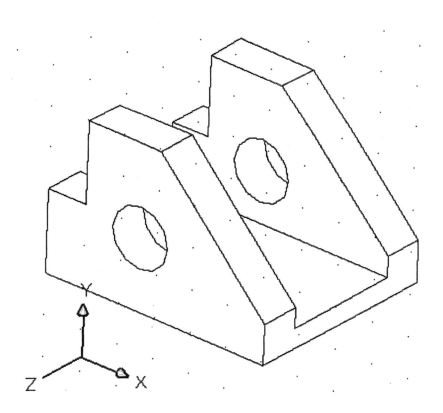

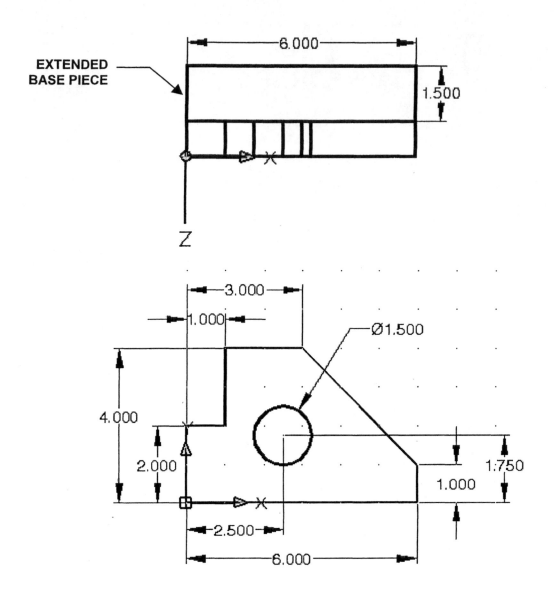

EXTENDED BASE PIECE

6.000

1.500

Z

3.000

1.000

Ø1.500

4.000

2.000

1.750

1.000

2.500

6.000

CONSTRUCTION HINTS

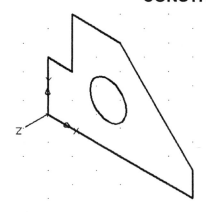

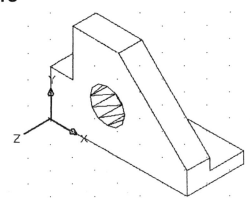

1. Draw 2D shape
2. Create Region
3. Subtract circle

4. Extrude Region
5. Add extended base piece
6. Union
7. Now **MIRROR3D**

17-8

EXERCISE 17B
ROTATE 3D

1. Open **EX-17A** (Figure 1)
2. Rotate the solid model to appear like Figure 2
3. Save as **EX-17B**

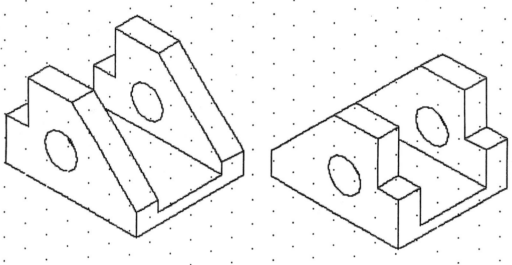

Figure 1   **Figure 2**

EXERCISE 17C
ALIGN

1. Open **My Decimal Setup**
2. Select the **Model tab**
3. Draw the 3 solid objects shown below in Figure 1
   (Use the dimensions shown)
4. Using the **ALIGN** command, assemble the objects as shown in Figure 2.
5. Save as **EX-17C**

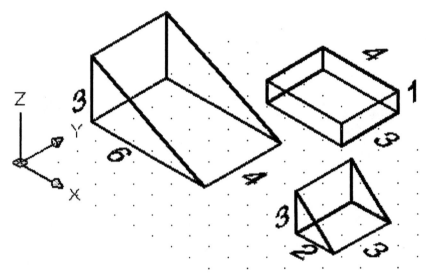

Figure 1

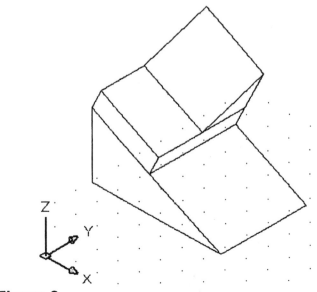

Figure 2

EXERCISE 17D
2D ARRAY

1. Open **My Decimal Setup**
2. Select the **Model tab**
3. Draw the box shown below in Figure 1 (Use the dimensions in Figure 2)
4. Try the **2D** Array command to draw the holes
5. Save as **EX-17D**

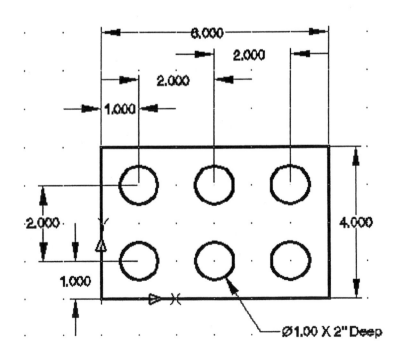

Figure 1

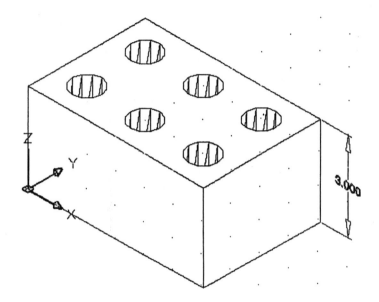

Figure 2

EXERCISE 17E
3D ARRAY - RECTANGULAR

1. Open **My Decimal Setup**
2. Select the **Model tab**
3. Draw the solid objects shown below in Figure 1
   (Use the dimensions shown at the bottom of the page)
4. Try the <u>3D Array</u> command to create the solid model in Figure 2.
5. Save as **EX-17E**

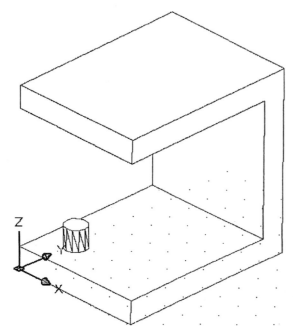

Figure 1

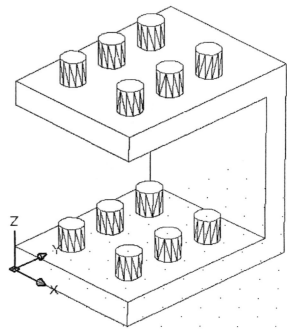

Figure 2

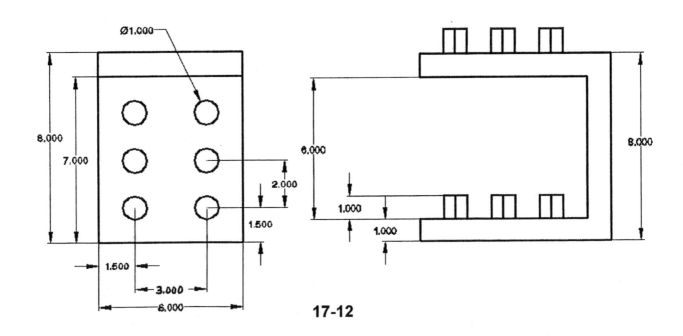

EXERCISE 17F
3D ARRAY - POLAR

1. Open **My Decimal Setup**
2. Select the **Model tab**
3. Create the solid cylinder with 8 holes shown below
    (Dimensions and construction hints on page 17-14)
4. Save as **EX-17F**

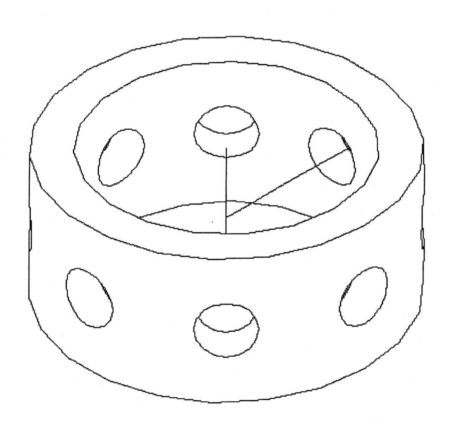

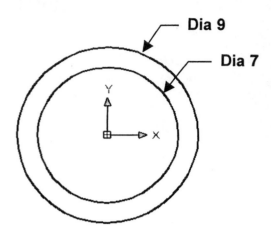

Dia 9

Dia 7

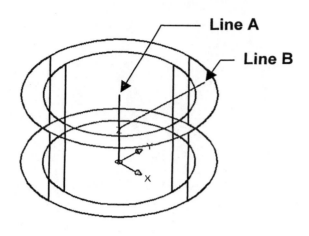

Line A

Line B

1. Draw 2 cylinders & subtract

2. Draw Line **A** <u>from</u> 0,0,0 <u>to</u> 0,0,4

3. Draw Line **B** <u>from</u> 0,0,2 <u>to</u> @0, 5.5, 0

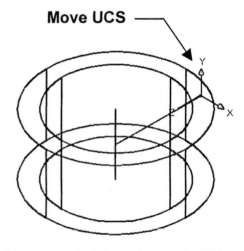

Move UCS

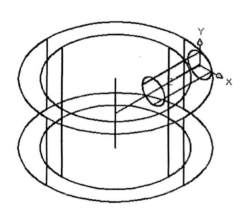

4. Move the UCS to the end of Line **B** and Rotate 90 degrees as shown

5. Draw the small 1.50 dia X 3 lg. cylinder Center point = 0,0,0

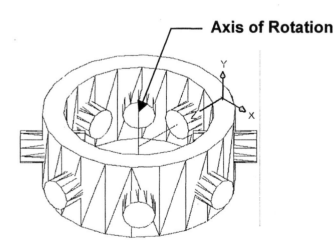

Axis of Rotation

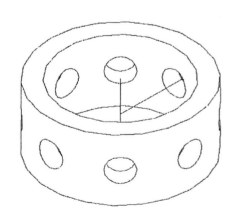

6. 3D Array the small cylinder (Axis of Rotation is Line **A)**

6. Subtract the small cylinders

LEARNING OBJECTIVES

After completing this lesson, you will be able to:

1. Understand 4 Solidedit commands.
   Extrude faces, Move faces, Offset faces and Delete faces

LESSON 18

SOLIDEDIT

AutoCAD has many editing commands that allow you to edit existing solids. In this Lesson four of the more frequently used editing commands will be discussed. They are: Extrude faces, Move faces, Offset faces and Delete faces.

Selecting a face to edit.
Selecting the correct face to edit is very important. To select a face, place your cursor on an <u>open area</u> of the face (surface). Do not select an edge. (Consider turning off Osnap. It may prevent you from placing the cursor in the correct location)
The edges of the face will become dashed to indicate which face has been selected.
(If the preferred face is under another face, left click again. Use "Remove" to de-select a face.

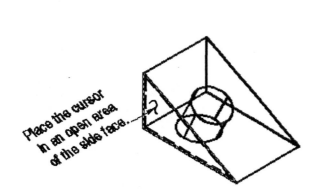

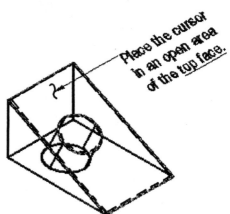

Positive vs. Negative value
<u>Positive</u> value <u>adds material</u>. <u>Negative</u> value <u>subtracts material</u>.
This sounds basic but you need to think about this. For example, if you need to <u>increase </u>the size of a <u>hole</u> you will enter a <u>negative</u> value because when the hole increases in size material is <u>subtracted</u> from the solid model.

ERROR MESSAGES
Sometimes, when using these commands, AutoCAD gets a little confused. If AutoCAD gets confused it stops working. If this happens you usually have to reboot the computer. So it is always a good idea to "**Save your drawing**" before attempting the Solidedit commands.

Also, AutoCAD may display an error message. The following are explanations for a few of the AutoCAD error messages.

"Invalid face/edge (or edge/edge, or face/face, etc. intersection"
You have selected a face that may interfere with another face after the operation.

"Gap cannot be filled"
If you try to delete a large face. Fill and Union instead if you get this message.

"Modeling error: [some command] will fail to produce a valid ACIS solid
AutoCAD is really confused and will not complete the command. Give up and find another method to edit your solid.

EXTRUDE FACES

The **Extrude faces** command adds or subtracts material perpendicular to the selected face. You will be prompted for a **height**. The height will always be perpendicular to the face and is not affected by the position of the UCS.
A Positive value will add material and a **Negative** value will remove material.

1. Select the **Extrude Faces** command using one of the following:

    TYPE = solidedit
    PULLDOWN = MODIFY / SOLIDS EDITING / EXTRUDE FACES
    TOOLBAR = SOLID EDITING ⟦⬚⟧ **(Note: the button looks just like the**
    Extrude command button but it is on a
    different toolbar)

2. Select faces or [Undo/Remove]: *select a face*

3. Select faces or [Undo/Remove/ALL]: *select more faces or <enter>*

4. Specify height of extrusion or [Path]: *type height (positive or negative)*

5. Specify angle of taper for extrusion <0>: *<enter>*

    Solid validation started.
    Solid validation completed.

6. Enter a face editing option
    [Extrude/Move/Rotate/Offset/Taper/Delete/Copy/coLor/Undo/eXit] <eXit>:*<enter>*

    Solids editing automatic checking: SOLIDCHECK=1

7. Enter a solids editing option [Face/Edge/Body/Undo/eXit] <eXit>: *<enter>*

Example:

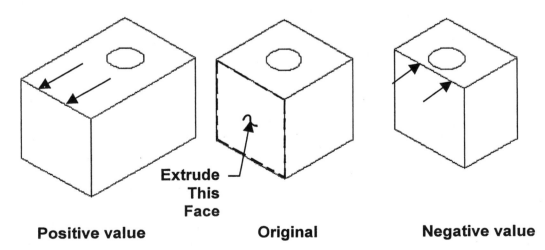

| Positive value | Original | Negative value |

MOVE FACES

The **Move Faces** command allows you to move a negative space such as a hole or an open space. You may also add or subtract material. Consider this the "Stretch" command for 3D solids.

1. Select the **Move Faces** command using one of the following:

   TYPE = solidedit
   PULLDOWN = MODIFY / SOLIDS EDITING / MOVE FACES
   TOOLBAR = SOLID EDITING

2. Select faces or [Undo/Remove]: ***select a face***

3. Select faces or [Undo/Remove/ALL]: ***select other faces or <enter>***

4. Specify a base point or displacement: ***select a base point***

5. Specify a second point of displacement: ***type relative coordinates or DDE***

   Solid validation started.
   Solid validation completed.

6. Enter a face editing option
   [Extrude/Move/Rotate/Offset/Taper/Delete/Copy/coLor/Undo/eXit] <eXit>: ***<enter>***

7. Solids editing automatic checking: SOLIDCHECK=1
   Enter a solids editing option [Face/Edge/Body/Undo/eXit] <eXit>:***<enter>***

   EXAMPLE:

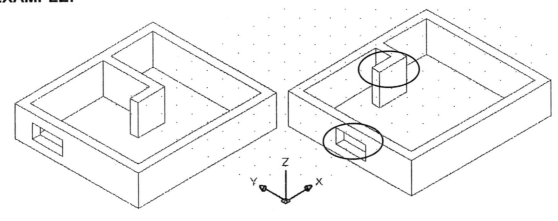

   Original   **The wall and Window moved**

OFFSET FACES

The **Offset Faces** command adds or removes material also. It is especially helpful when editing the size of holes.

Remember, a <u>Negative</u> value <u>removes material</u> from the solid, so the <u>hole</u> will get <u>larger</u>. A <u>Positive</u> value <u>adds material</u> to the solid, so the hole will get <u>smaller</u>.

1. Select the **Offset Faces** command using one of the following:

    TYPE = solidedit
    PULLDOWN = MODIFY / SOLIDS EDITING / OFFSET FACES
    TOOLBAR = SOLID EDITING

2. Select faces or [Undo/Remove]:*select a face*

Note: If you select a hole only part of the circle will dash. That's OK, the entire hole is actually selected. This is a minor display bug.

3. Select faces or [Undo/Remove/ALL]:*select more faces or <enter>*

4. Specify the offset distance: *type the offset distance*

    Solid validation started.
    Solid validation completed.

5. Enter a face editing option
    [Extrude/Move/Rotate/Offset/Taper/Delete/Copy/coLor/Undo/eXit] <eXit>: *<enter>*

    Solids editing automatic checking: SOLIDCHECK=1

6. Enter a solids editing option [Face/Edge/Body/Undo/eXit] <eXit>:*<enter>*

EXAMPLE:

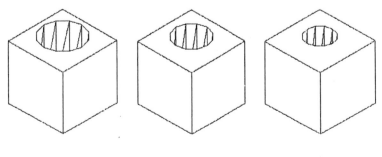

Negative value   **Original**   **Positive value**

DELETE FACES

The **Delete Faces** command removes a face. Use this command to delete holes.

1. Select the **Delete Faces** command using one of the following:

    TYPE = solidedit
    PULLDOWN = MODIFY / SOLIDS EDITING / DELETE FACES
    TOOLBAR = SOLID EDITING

2. Select faces or [Undo/Remove]: *select a face (hole)*

3. Select faces or [Undo/Remove/ALL]: *select another face or <enter>*

    Solid validation started.
    Solid validation completed.

4. Enter a face editing option
    [Extrude/Move/Rotate/Offset/Taper/Delete/Copy/coLor/Undo/eXit] <eXit>: *<enter>*

    Solids editing automatic checking: SOLIDCHECK=1

5. Enter a solids editing option [Face/Edge/Body/Undo/eXit] <eXit>:*<enter>*

EXAMPLE:

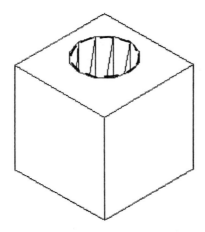
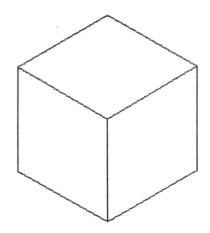

Original with hole   **Hole deleted**

EXERCISE 18A
EXTRUDE FACES

1. Open **EX-17A**
2. Select the **Model tab**
3. Edit the solid model as shown.
4. Save as **EX-18A**

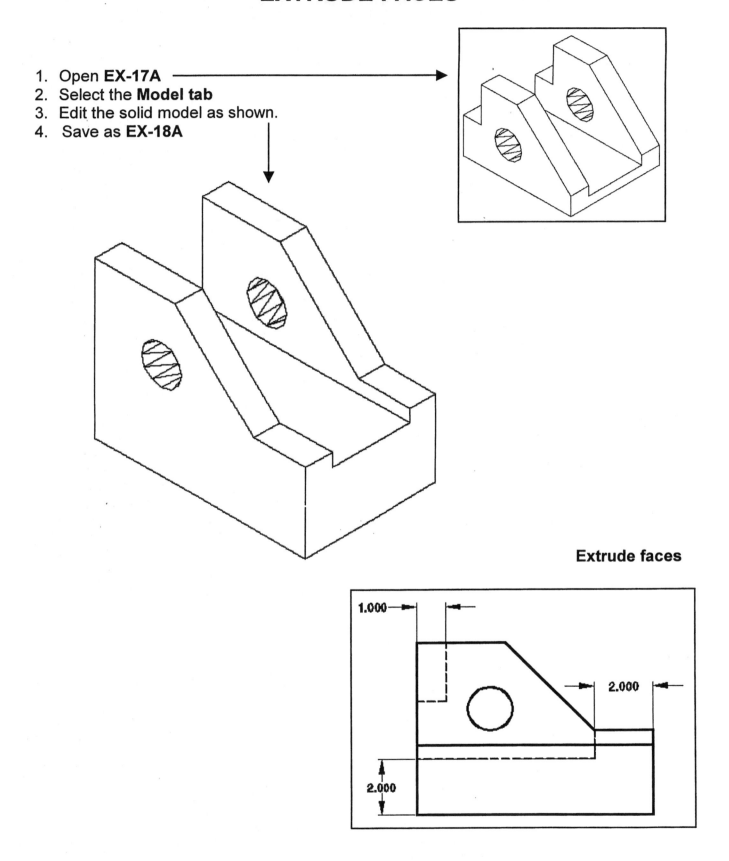

Extrude faces

1.000

2.000

2.000

EXERCISE 18B
MOVE FACES

1. Open **EX-16E**
2. Edit the solid model as shown below.
3. Refer to page 18-9 for instructions for removing an opening.
4. Save as **EX-18B**

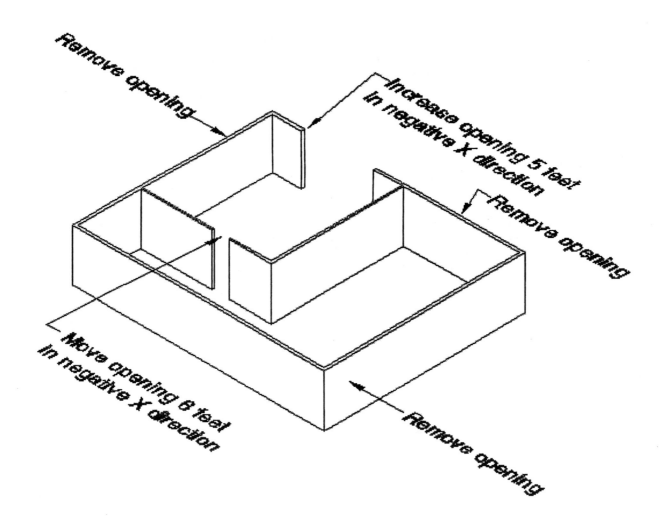

HOW TO CLOSE OPENING.

1. Select the **MOVE FACES** command.

2. Select the face.

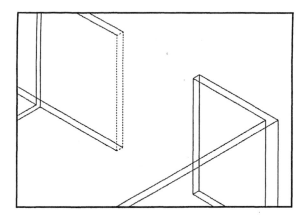

3. Specify the basepoint

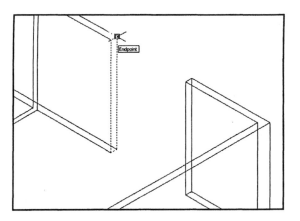

4. Specify the new location.

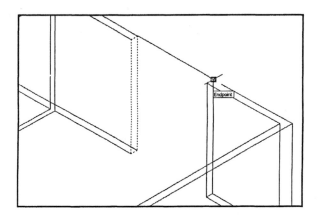

5. Union

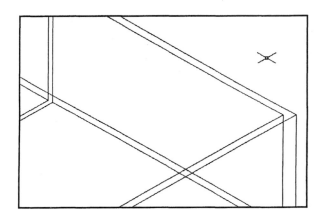

EXERCISE 18C
OFFSET FACES

1. Open **My Decimal Setup**
2. Draw the solid shown in Figure 1. (Dimensions on page 18-11)
3. Save as **EX-18C1**

FIGURE 1

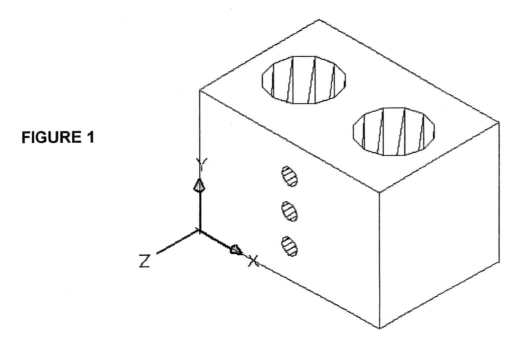

4. Edit the holes per dimensions on page 18-11.
5. Save as **EX-18C2**

FIGURE 2

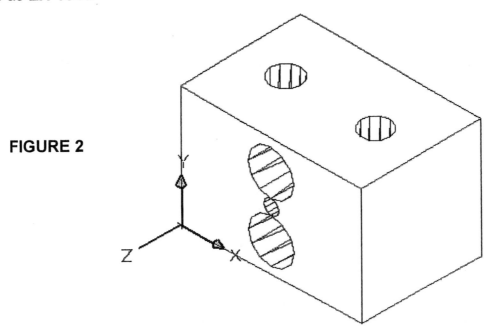

DIMENSIONS FOR FIGURE 1

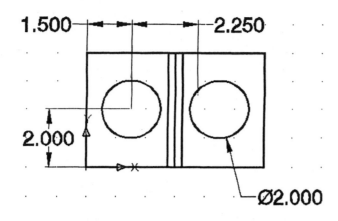

1.500 ⟵⟶ 2.250

2.000

Ø2.000

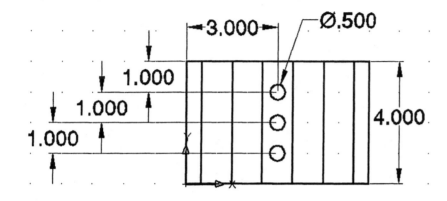

3.000 ⟶ Ø.500

1.000

1.000

1.000

4.000

DIMENSIONS FOR FIGURE 2

Ø1.000

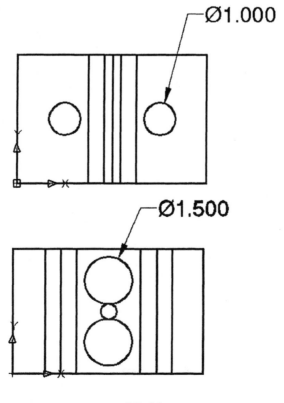

Ø1.500

EXERCISE 18D
DELETE FACES

1. Open **EX-18A**
2. Delete the holes as shown in Figure 2.
3. Save as **EX-18D**

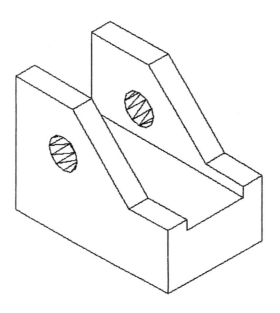

Figure 1

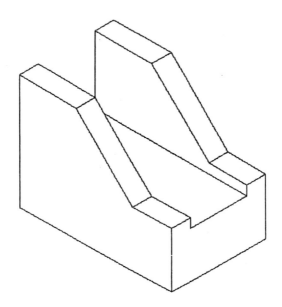

Figure 2

LEARNING OBJECTIVES

After completing this lesson, you will be able to:

1. Create a solid object by revolving a 2D shape
2. Slice a solid object into 2 segments
3. Create a section view through a solid object.

LESSON 19

REVOLVE

The **REVOLVE** command allows you to create a solid by revolving a closed shape about an axis or an object. The closed shape can be revolved from 1 to 360 degrees. You select the axis to revolve about, the solid to be revolved and enter the angle of revolution.

Selecting the axis of revolution
The axis of revolution can be the X or Y axis. It may also be an object such as a line. The axis may be on the closed shape or not.

HOW TO USE THE REVOLVE COMMAND
1. Select the **REVOLVE** command using one of the following:

> **TYPE = REV or REVOLVE**
> **PULLDOWN = DRAW / SOLIDS / REVOLVE**
> **TOOLBAR = SOLIDS**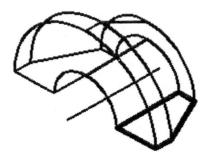

2. Select objects: *select the closed shape or region*

3. Select objects: *select more or <enter>*

4. Specify start point for axis of revolution or
   define axis by [Object/X (axis)/Y (axis)]: *select the X, Y or Object option*

5. Select an object: *select the axis of object to revolve about*

6. Specify angle of revolution <360>: *type the angle (positive or negative determines direction of rotation.*

Examples of axis of revolution (180 degrees)

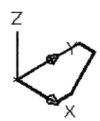

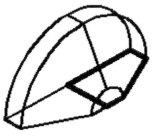

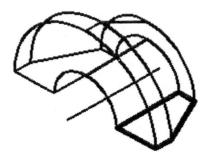

X AXIS   **Y AXIS**   **OBJECT**

SLICE

The **SLICE** command allows you to make a knife cut through a solid. You specify the location of the slice by specifying the plane. After the solid has been sliced, you determine which portion of the slice you want removed. You may also keep both portions but they are still separate.

There are 2 methods to specify the cutting plane.

Method 1.
1. Define the cutting plane by selecting 3 points on the object.

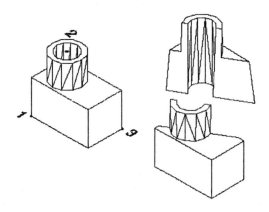

Method 2.
1. Move the UCS to the desired cutting plane location.
2. Select the XY, YZ or ZX plane. (The cut will be made along the selected plane)

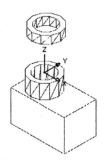   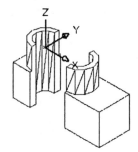   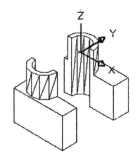

XY PLANE   **YZ PLANE**   **ZX PLANE**

HOW TO USE THE SLICE COMMAND
1. Select the **SLICE** command using one of the following:

> **TYPE = SL or SLICE**
> **PULLDOWN = DRAW / SOLIDS / SLICE**
> **TOOLBAR = SOLIDS**

2. Select objects: *select the solid object*
3. Select objects: *<enter>*
4. Specify first point on slicing plane by [Object/Zaxis/View/XY/YZ/ZX/3points] <3points>: *select the option preferred*

SECTION

The **SECTION** command process is very similar to the Slice command. But the Section command does not actually separate the solid. It creates a 2D region from the plane that you specify. The 2D region will be created on the current layer. The 2D Region can then be moved, extruded, revolved, copied, moved, etc.

To specify the section plane refer to the Slice methods on page 19-3.

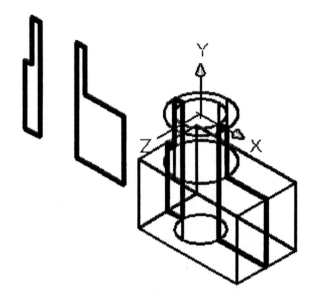

HOW TO USE THE SECTION COMMAND
1. Select the **SECTION** command using one of the following:

> **TYPE = SEC or SECTION**
> **PULLDOWN = DRAW / SOLIDS / SECTION**
> **TOOLBAR = SOLIDS**

2. Select objects: *select the solid object*
3. Select objects: *<enter>*
4. Specify first point on section plane by [Object/Zaxis/View/XY/YZ/ZX/3points] <3points>: *select the option preferred*

HOW TO CREATE A HATCHED SECTION VIEW
1. Create a section (2D region) as described above.
2. Move or copy the region to a new location.
3. Explode if necessary to create separate regions.
4. Use the Hatch command to create section lines.
   (Note: The area to be hatched must be parallel to the XY plane or you will receive an error message. *My suggestion is, if you intend to hatch a section, use the XY plane method for defining the plane.)*
5. Draw any connecting lines to complete the view.

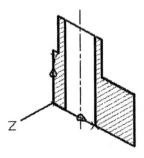

EXERCISE 19A
REVOLVE

1. Open **My Decimal Setup**
2. Select the **Model tab**
3. Draw the 2D shape shown below.
4. Create 2 copies (3 total)
5. Revolve in the X axis, Y axis and about an object, as shown on the next page.
6. Save as **EX-19A**

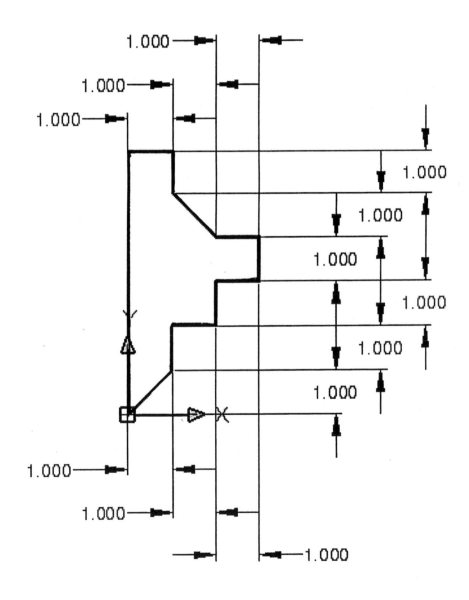

ORIGINAL SHAPE

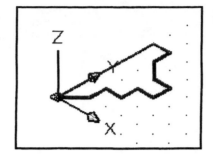

ROTATE 180 DEGREES ABOUT THE X AXIS

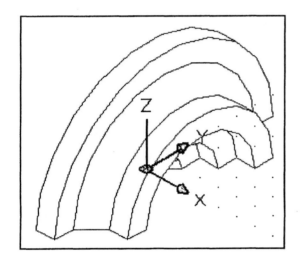

ROTATE –90 DEGREES ABOUT THE Y AXIS

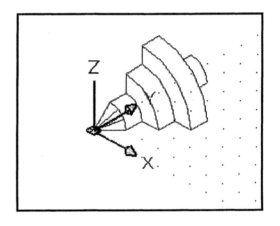

ROTATE 360 DEGREES ABOUT AN OBJECT

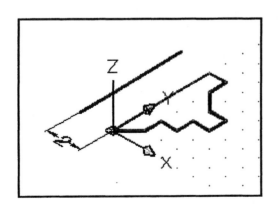

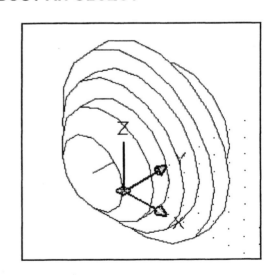

EXERCISE 19B
SLICE

1. Open **My Decimal Setup**
2. Select the **Model tab**
3. Draw the 3D solid shown in the upper right view **FIGURE 1**. (The other 3 views are for your information only)
4. Save as **EX-19B1**
5. Split the solid object as shown.
6. **SLICE** the solid as shown in **FIGURE 2.**
7. Save as **EX-19B2**

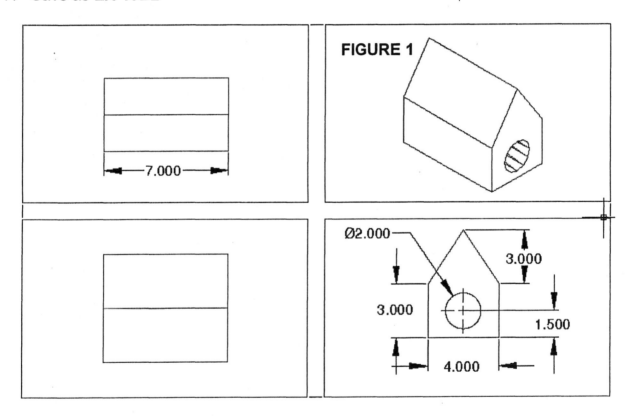

FIGURE 1

7.000

Ø2.000

3.000

3.000

1.500

4.000

FIGURE 2

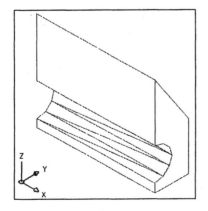

EXERCISE 19C
SECTION

1. Open **19B1**
2. Select the **Model tab**
3. Create a **SECTION** as shown.
4. The new section should be on the "**Section**" layer.
5. The linetype should be **Continuous**.
6. Save as **EX-19C**

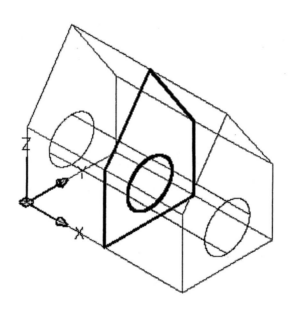

EXERCISE 19D
HATCH THE SECTION

1. Open **19C**
2. Move the section (2D region) away from the solid model.
3. **HATCH** the section. (Refer to page 19-4)
4. Place the hatch lines on the "Hatch" layer.
5. Hatch specs: Angle = 0 Scale = 4
6. Save as **EX-19D**

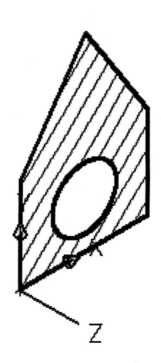

Z

Notes:

LEARNING OBJECTIVES

After completing this lesson, you will be able to:

1. Create a multiple view plot setup
2. Control the hidden line display for plotting
3. Shade one of the views
4. Dimension a multiple view display

LESSON 20

PLOTTING MULTIPLE VIEWS

AutoCAD has many commands to create a multiview layout, such as SOLVIEW, SOLDRAW, SOLPROF and DXB plot. You may consider researching these. But in this lesson I will show you 5 quick and simple steps to develop a multiview layout to use for plotting.

STEP 1. SELECTING THE PLOTTER / PRINTER AND PAPER SIZE
1. Open the solid model drawing
2. Select an unused Layout tab. If all Layout tabs have been used, create a new Layout tab. Name it "multiview." (Refer to page 4-17, B-1 for instructions)
3. Select a "Plotter / Printer", "Plot Style table" and "Paper Size".
4. Select **OK**

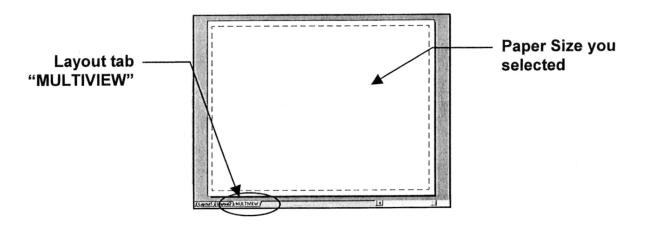

Layout tab "MULTIVIEW"

Paper Size you selected

STEP 2. CREATE VIEWPORTS
5. Select **VIEW / VIEWPORTS / NEW VIEWPORTS**
   a. Select **FOUR EQUAL**
   b. Set **Viewport Spacing to: .50**
   c. Set **Setup to: 3D**
   d. Select **OK**
   e. Specify first corner or [Fit] <Fit>: **<enter>**

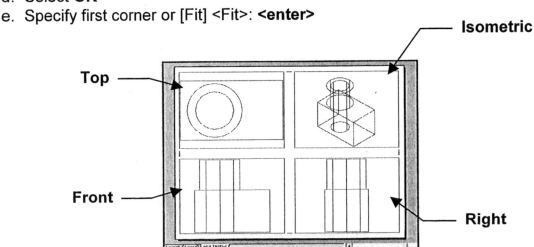

Isometric

Top

Front

Right

STEP 3. ADJUST VIEWPORT SCALE, PAN AND LOCK
6. Open the VIEWPORT toolbar and adjust the scale of each Viewport.
7. Pan the object in each viewport to the desired location. (Do not use Zoom)
8. LOCK each viewport

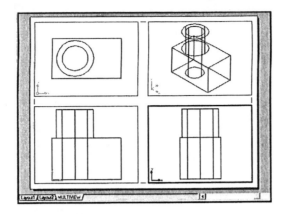

STEP 4. SELECT THE PLOT APPEARANCE
9. Make sure you are in Paperspace
10. Select **ALL** the viewports.
11. Right click and select "PROPERTIES" from the menu
12. Find the "MISC" category and select **SHADE PLOT = HIDDEN**
13. Press the **ESC** key and close the Properties palette.

STEP 5. PLOT
14. Select **FILE / PLOT**
15. Select **PLOT DEVICE** tab.
    a. Confirm that the correct Plotter / Printer is displayed.
    b. Confirm that the correct Plot Style table is displayed.
16. Select **PLOT SETTING** tab.
    a. Confirm that the correct paper size is displayed.
    b. Select **PLOT AREA = EXTENTS.**
    c. Select **PLOT SCALE = SCALED TO FIT**
    d. Select **PLOT OFFSET = CENTER THE PLOT**
17. Select the **FULL PREVIEW** button.
    a. Press **<enter>** or **<esc>** if preview is correct. (If not, recheck the settings)
    b. Select the **OK** button.

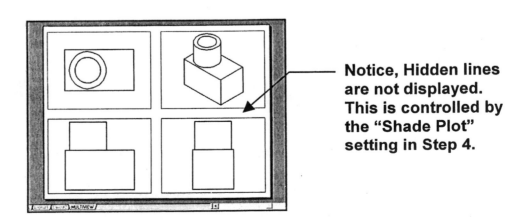

Notice, Hidden lines are not displayed. This is controlled by the "Shade Plot" setting in Step 4.

20-3

How to control the hidden line plot

Option 1. Invisible hidden lines
In Step 4, on the previous page, you set the SHADE PLOT to HIDDEN. The "Hidden" setting displays the hidden lines, in the Full Preview, as "invisible" for plotting. They are not actually removed. When you return to the drawing, the hidden lines are still there.

Option 2. Dashed hidden lines
You may want to plot the drawing with hidden lines shown. To accomplish this you must change the **OBSCUREDLTYPE** variable setting.

1. On the command line type **obscuredltype <enter>**
2. Enter new value for OBSCUREDLTYPE <0>: *type 2 <enter>*

The Obscured Linetype variable controls the appearance of the hidden lines. There are 11 choices as follows; **0 Off**, 1 Solid, **2 Dashed**, 3 Dotted, 4 Short Dash, 5 Medium Dash, 6 Long Dash, 7 Double Short Dash, 8 Double Medium Dash, 9 Double Long Dash, 10 Medium Long Dash and 11 Sparse Dot.

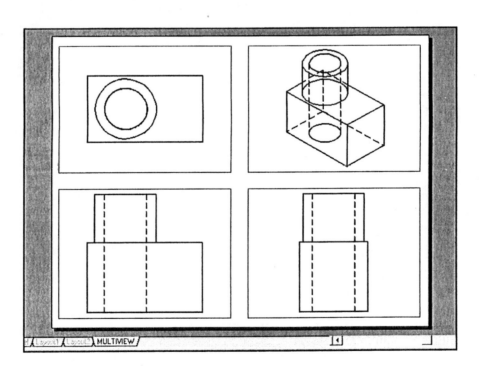

Option 3. Hidden lines a different color.
You may want the hidden lines to be a different color. To accomplish this you must change the **OBSCUREDCOLOR** setting.

1. On the command line type **obscuredcolor <enter>**
2. Enter new value for OBSCUREDCOLOR <0>: *type the color number <enter>*

To print in color you must select a "color dependent plot style table" with colors assigned. Refer to Appendix-C for instructions.

How to plot with a shaded view

1. First shade the model inside of the viewport.
2. Now return to Paperspace
3. Select the shaded viewport frame.
4. Right click and select **PROPERTIES** from the menu
5. Find the **MISC** category and select **SHADE PLOT = <u>AS DISPLAYED</u>**
6. Press the **ESC** key and close the Properties palette.
7. Follow the instructions in Step 5 on page 20-3 to plot.

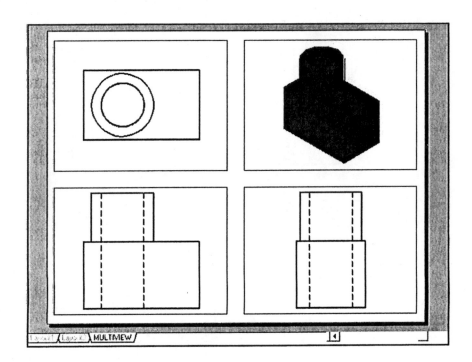

DIMENSIONING MULTIPLE VIEWS

It is important to dimension in Paperspace.

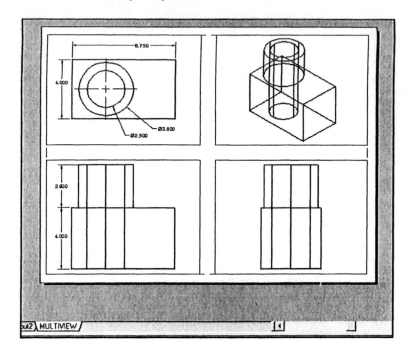

If you dimension in modelspace, each dimension will be visible in all viewports.

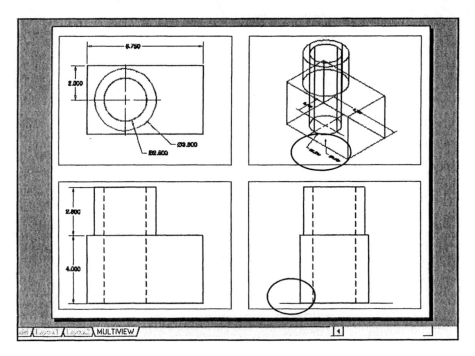

EXERCISE 20A
MULTIVIEW PLOT SETUP

1. Open **EX-14E**

2. Create a **new** Layout tab.

3. Select the **HP 4MV** plotter and **11 X 17** paper.

4. Create a **multiview plot setup** as shown below.
   (Refer to the instructions on page 20-2 to 20-3)

5. Adjust the scale in each viewport to 3/8 and PAN.

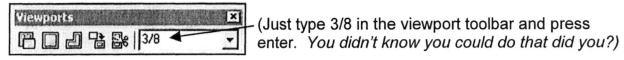 (Just type 3/8 in the viewport toolbar and press enter. *You didn't know you could do that did you?*)

6. Save as **EX-20A**

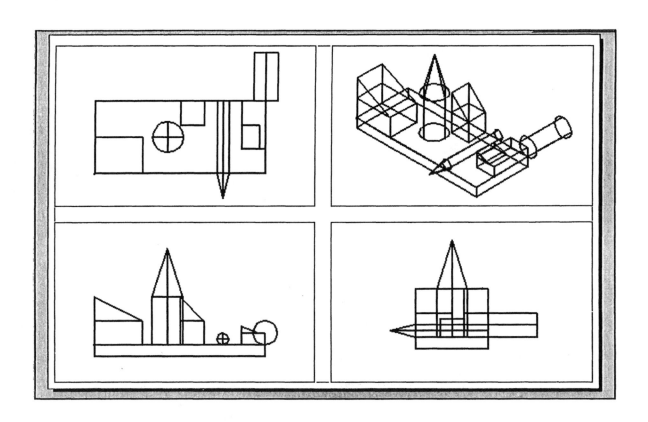

EXERCISE 20B
INVISIBLE HIDDEN LINES

1. Open **EX-20A**

2. Set the **SHADE PLOT** to **HIDDEN** in all of the viewports.
   (Refer to page 20-3 for instructions)

3. Set the **DISPSILH** to 1.

4. Select **FILE / PLOT**.

5. Select **FULL PREVIEW** button. Does your drawing look like the drawing below?

6. Save as **EX-20B**

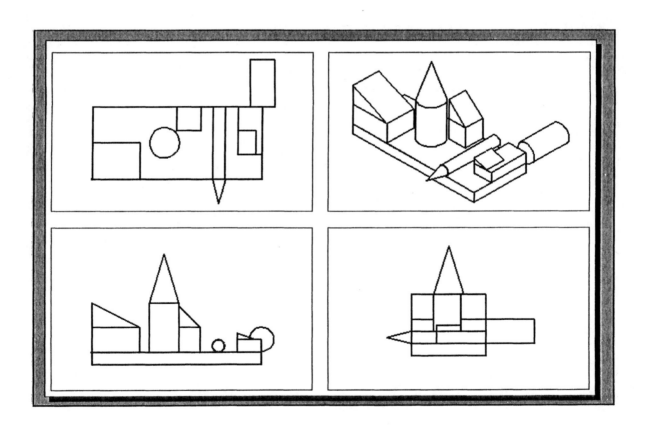

EXERCISE 20C
VISIBLE DASHED HIDDEN LINES

1. Open **EX-20B**

2. Set the **OBSCUREDLTYPE** variable to display dashed hidden lines.
   (Refer to page 20-4 for instructions)

3. Select **FILE / PLOT**.

4. Select **FULL PREVIEW** button. Does your drawing look like the drawing below?

5. Save as **EX-20C**

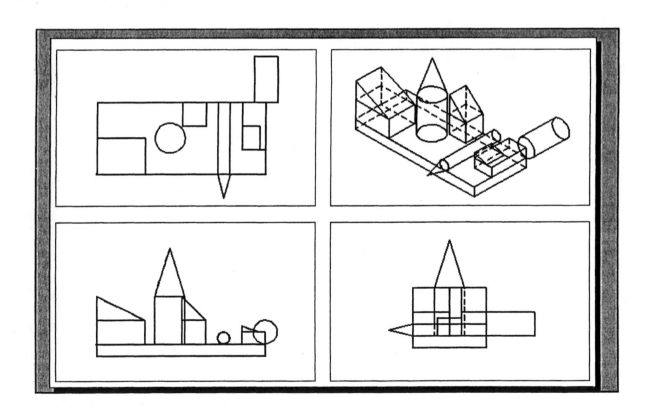

EXERCISE 20D
DISPLAY A SHADED VIEW

1. Open **EX-20B (not 20C)**

2. Shade the isometric view.

3. Set the **SHADEPLOT** setting to display the shaded view when plotted.
    (Refer to page 20-5 for instructions)

4.   Select **FILE / PLOT**.

5.   Select **FULL PREVIEW** button. Does your drawing look like the drawing below?

6. Save as **EX-20D**

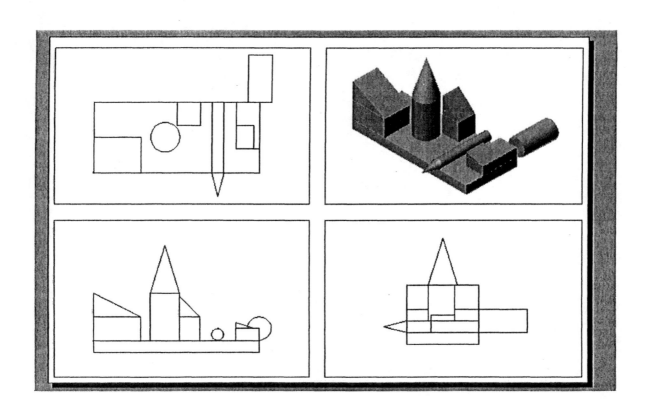

EXERCISE 20E

DIMENSIONING

1. Open **EX-20B (not 20D)**

2. Dimension as shown below.
   (Refer to page 20-6 for instructions.)

3. Save as **EX-20E**

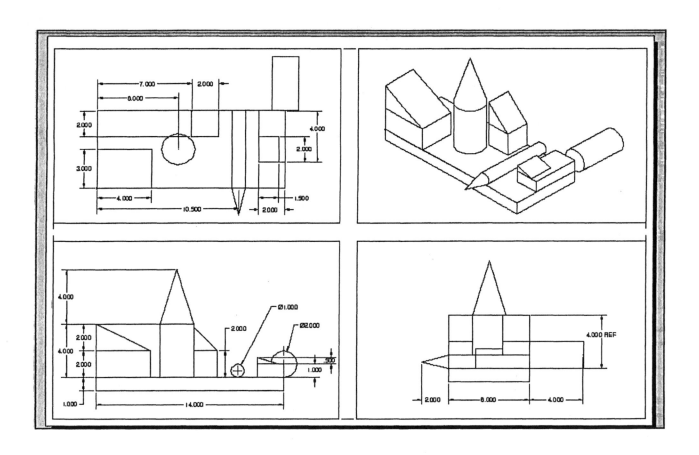

NOTES:

ARCHITECTURE

ARCHITECTURAL SYMBOL LIBRARY

When you are using a CAD system, you should make an effort to only draw an object once. If you need to duplicate the object, use a command such as: Copy, Array, Mirror or Block. This will make drawing with CAD more efficient.

In the following exercise, you will create a file full of architectural symbols that you will use often when creating an architectural drawing. You will create them once and then merely drag and drop them, from the DesignCenter, when needed. This will save you many hours in the future.

Save this library file as **Library** so it will be easy to find when using the DesignCenter or create a Library Palette.

1. Open **My Feet-Inches Setup**

2. Select the **Qtr equals foot** tab.

3. Draw each of the Symbol objects, shown on the following pages, actual size. Do not scale them.

4. Create an individual Block for each one using the **BLOCK** command.

5. Save this drawing as: **Library**

6. Plot the drawing, of your library symbols, for reference.
    The format is your choice.
    Use Page setup: **24 X 18 All Black**

Consider creating a library palette. Refer to page 10-11.

SYMBOLS	INSTRUCTIONS

SYMBOLS   **INSTRUCTIONS**

1

Ø8"

2"

EXTEND 2"
BEYOND
CIRCLE

1'

DUPLEX CONVENIENCE OUTLET

LAYER = ELECTRICAL
CIRCLE = COLOR: RED
3 HORIZONTAL LINES = COLOR: BLUE

2

GFI

GROUNDED DUPLEX OUTLET

DIMENSIONS, LAYER AND COLOR THE SAME
AS SYMBOL NUMBER 1

GFI TEXT = COLOR= BLUE   HEIGHT= 3"

POSITION TEXT APPROXIMATELY AS SHOWN

3

GFI

WP

GROUNDED WEATHER PROOF OUTLET

4

220V

220 VOLT OUTLET

3 HORIZONTAL LINES = COLOR: BLUE

5

8"

4"

WALL MOUNTED FIXTURE W/INCANDESCENT LAMP

DIMENSIONS NOT SHOWN ARE THE SAME AS SYMBOL
NUMBER 1.

LAYER = ELECTRICAL

6

2"

CEILING MOUNTED FIXTURE W/FLUORESCENT LAMP

7

SUSPENDED FIXTURE W/INCANDESCENT LAMP

8

2"

RECESSED FIXTURE W/INCANDESCENT LAMP

ARCH-3

SYMBOLS		INSTRUCTIONS

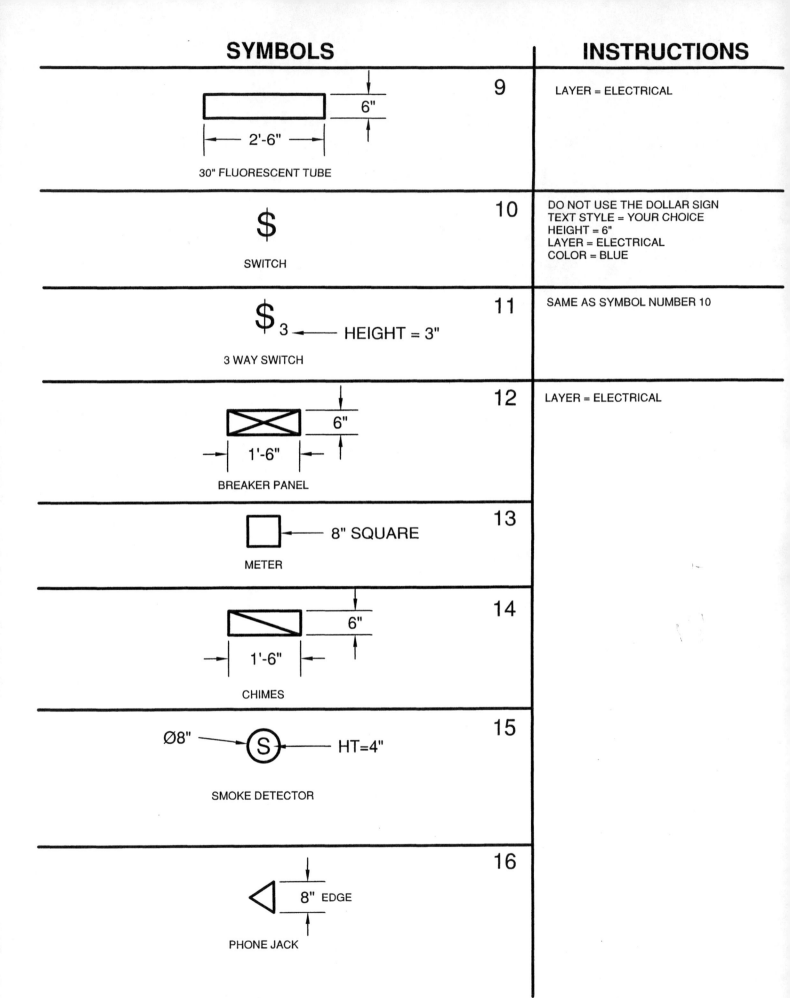

SYMBOLS

INSTRUCTIONS

9

30" FLUORESCENT TUBE
(2'-6" × 6")

LAYER = ELECTRICAL

10

$

SWITCH

DO NOT USE THE DOLLAR SIGN
TEXT STYLE = YOUR CHOICE
HEIGHT = 6"
LAYER = ELECTRICAL
COLOR = BLUE

11

$3 ← HEIGHT = 3"

3 WAY SWITCH

SAME AS SYMBOL NUMBER 10

12

BREAKER PANEL
(1'-6" × 6")

LAYER = ELECTRICAL

13

8" SQUARE

METER

14

CHIMES
(1'-6" × 6")

15

Ø8" → (S) ← HT=4"

SMOKE DETECTOR

16

8" EDGE

PHONE JACK

ARCH-4

SYMBOLS

INSTRUCTIONS

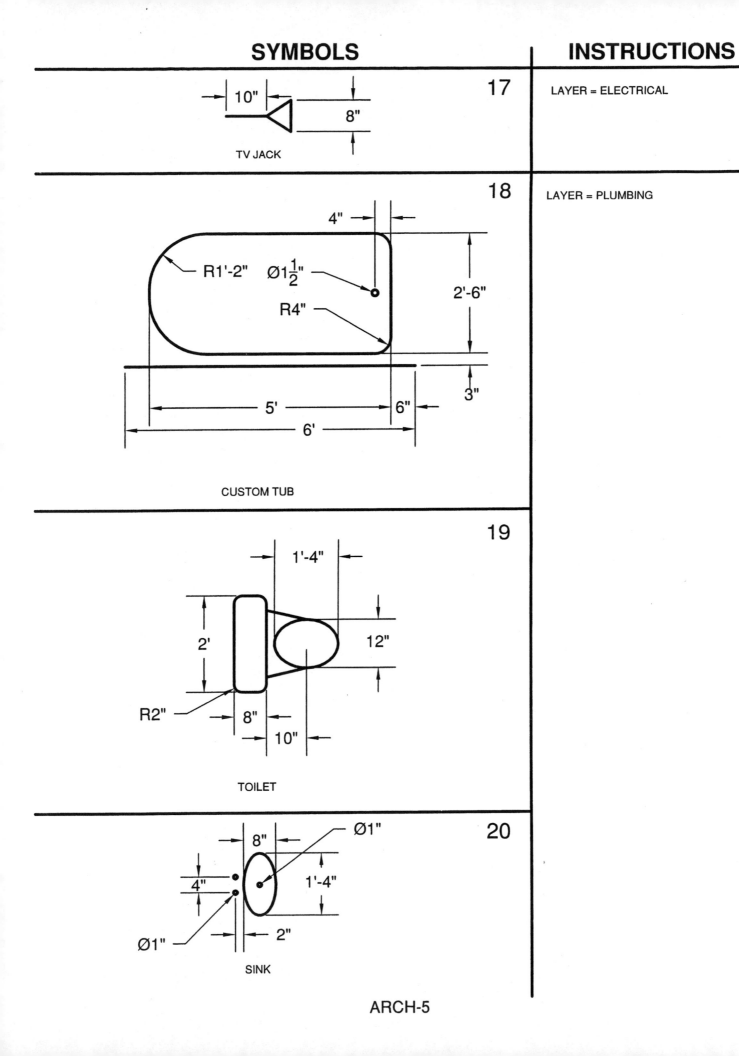

17

LAYER = ELECTRICAL

10"

8"

TV JACK

18

LAYER = PLUMBING

4"

R1'-2"   Ø1½"

R4"

2'-6"

3"

5'   6"

6'

CUSTOM TUB

19

1'-4"

2'

12"

R2"   8"

10"

TOILET

20

8"   Ø1"

4"

1'-4"

Ø1"   2"

SINK

ARCH-5

SYMBOLS

INSTRUCTIONS

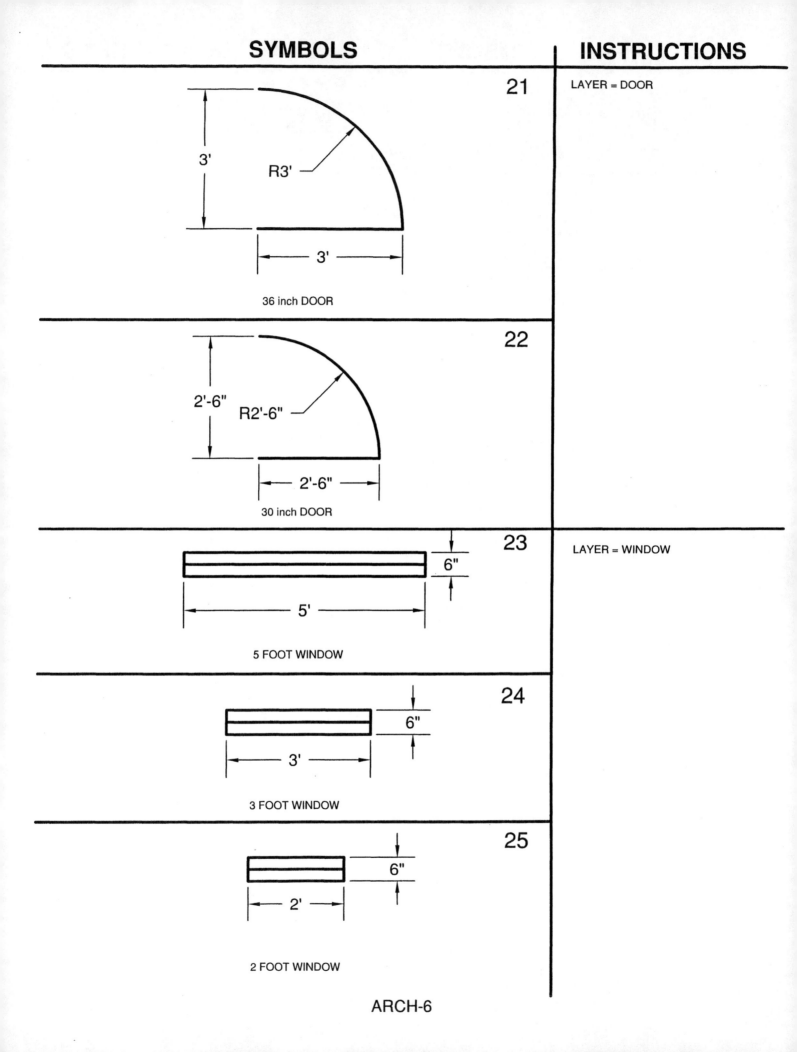

21

LAYER = DOOR

3'

R3'

3'

36 inch DOOR

22

2'-6"

R2'-6"

2'-6"

30 inch DOOR

23

LAYER = WINDOW

6"

5'

5 FOOT WINDOW

24

6"

3'

3 FOOT WINDOW

25

6"

2'

2 FOOT WINDOW

ARCH-6

SYMBOLS	INSTRUCTIONS

26

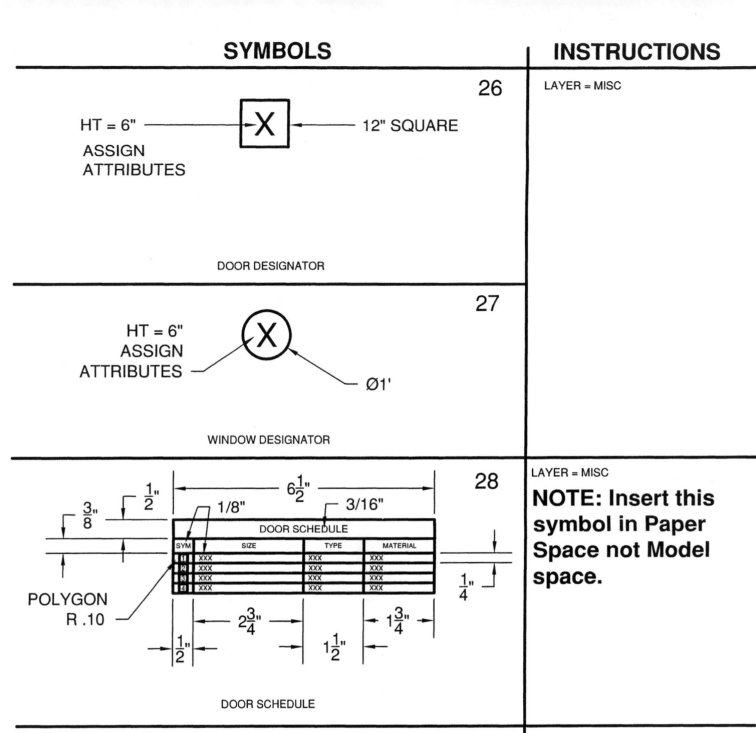

HT = 6" ———————— 12" SQUARE
ASSIGN
ATTRIBUTES

DOOR DESIGNATOR

LAYER = MISC

27

HT = 6"
ASSIGN
ATTRIBUTES ——————— Ø1'

WINDOW DESIGNATOR

28

POLYGON
R .10

DOOR SCHEDULE			
SYM	SIZE	TYPE	MATERIAL
1	XXX	XXX	XXX
2	XXX	XXX	XXX
3	XXX	XXX	XXX
4	XXX	XXX	XXX

DOOR SCHEDULE

LAYER = MISC

NOTE: Insert this symbol in Paper Space not Model space.

29

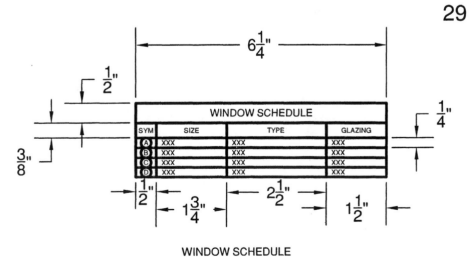

WINDOW SCHEDULE			
SYM	SIZE	TYPE	GLAZING
A	XXX	XXX	XXX
B	XXX	XXX	XXX
C	XXX	XXX	XXX
D	XXX	XXX	XXX

WINDOW SCHEDULE

LAYER = MISC

TEXT HT = SAME AS SYMBOL NUMBER 28

NOTE: Insert this symbol in Paper Space not Model space.

ARCH-7

ARCH-8

SITE PLAN

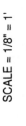

EXERCISE 158-1

SCALE = 1/8" = 1'

LOT 2 BLOCK A
PINE HILLS SUBDIVISION
SANTA ANA, CALIFORNIA

Do not use dimensions
for this length.
Use Single Line Text
Ht = 1/8" in Paper Space.

90'-0"

100'-0"

100'-0"

90'-0"

60'

34'

28'

16'

4'

20'

INSTRUCTIONS:

1. Open **MY Feet-Inches SETUP**
2. Select the **EIGHTH EQUALS FOOT** tab.
3. Draw the site plan above, in model space, full scale.
4. Dimension in Paper Space. Make sure True Associative dimensioning is On. (Dimassoc=2)
5. Design your own North symbol.
6. Fill in the information in the title block area.
7. Save as: **EX-158-1**
8. Plot using Page Setup: **24 X18 ALL BLACK**

EXERCISE 158-2

SCALE = 1/4" = 1'

LOT 2 BLOCK A
PINE HILLS SUBDIVISION
SANTA ANA, CALIFORNIA

INSERT IN PAPER SPACE
ON LAYER SYMBOL

Floor Plan Labels

LIVING/SLEEPING

BATH

NOOK

KITCHEN

Dimensions: 26', 6'-9", 3', 16'-3", 11', 6'-6", 1'-7", 5'-6", 5'-6", 2'-6", 2', 5', 9', 14', 7', 4', 6', 30', 6'-6", 6'-6", 6'-6", 6'-6"

DOOR SCHEDULE

SYM	SIZE	TYPE	MATERIAL
1	6'-0" X 6'-8"	WOOD SLIDER	1/4" POL PL
2	3'-0" X 6'-8"	PANEL	STAIN GRADE
3	2'-6" X 6'-8"	H.C. SLAB	STAIN GRADE
4	4'-0" X 6'-8"	H.C. SLAB	STAIN GRADE

WINDOW SCHEDULE

SYM	SIZE	TYPE	GLAZING
A	5'-0" X 4'-0"	WOOD FIXED	3/16" SHEET
B	3'-0" X 4'-0"	WOOD FIXED	3/16" SHEET
C	3'-0" X 3'-0"	WOOD FIXED	3/16" SHEET
D	2'-0" X 3'-0"	ALUMINUM SLIDER	3/16" SHEET

INSTRUCTIONS:

1. Open **MY FEET-INCHES SETUP**
2. Select the **QTR EQUALS FOOT** tab.
3. Draw the floor plan above, in model space, full scale. (Dimension in Paper Space, Dimassoc = 2)
4. Room title Ht. = 6"   Walls = 6" wide   Cabinets = 24" deep.
5. Insert the SCHEDULES (ON LAYER "SYMBOLS") using the DesignCenter and your Library dwg.
6. Fill in the information in the title block area.
7. Save as: **EX-158-2**
8. Plot using Page Setup: **24 X18 ALL BLACK**

LEGEND OF ELECTRICAL SYMBOLS

SYM	DESCRIPTION
	DUPLEX CONVENIENCE OUTLET
	GROUNDED DUPLEX OUTLET
	GROUNDED WEATHER PROOF OUTLET
220v	220 VOLT OUTLET
	WALL MTD. FIXT. W/ INCANDESCENT LAMP
	CEILING MTD. FIXT. W/INCANDESCENT LAMP
	SUSPENDED FIXT. W/INCANDESCENT LAMP
	RECESSSED FIXT. W/INCANDESCENT LAMP
	3'-0" FLUORESCENT TUBE
$	SWITCH
$₃	3 WAY SWITCH
	BREAKER PANEL
	METER
	CHIMES
S	SMOKE DETECTOR
	PHONE JACK
	T.V. JACK AND LEAD-IN

TEXT HT = 1/8" (IN PAPERSPACE)

5'

1'

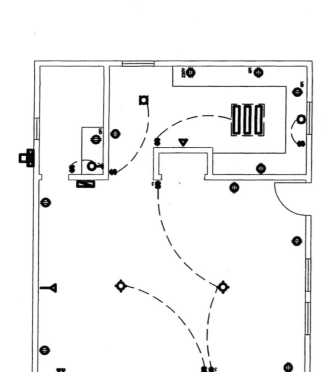

LOT 2 BLOCK A
PINE HILLS SUBDIVISION
SANTA ANA, CALIFORNIA

SCALE = 1/4" = 1'

EXERCISE 158-3

INSTRUCTIONS:

1. Open **158-2** and immediately save it as **158-3** for safety.
2. Select the **QTR EQUALS FOOT** tab.
3. Freeze Layers: Dimension, Misc, and Symbols.
4. Change the remaining Floor Plan to Color Magenta (Use Modify / Properties)
   Now the Floor Plan will plot light and the Electrical Symbols heavy.
5. Now insert the Electrical symbols (on layer Electrical)
6. Draw the "Legend Of Electrical Symbols" on Layer BORDER and TEXT-LIT in paperspace.
7. Fill in the information in the title block area.
6. Save as: **EX-158-3** and Plot using Page Setup: **24 x 18 ALL BLACK**

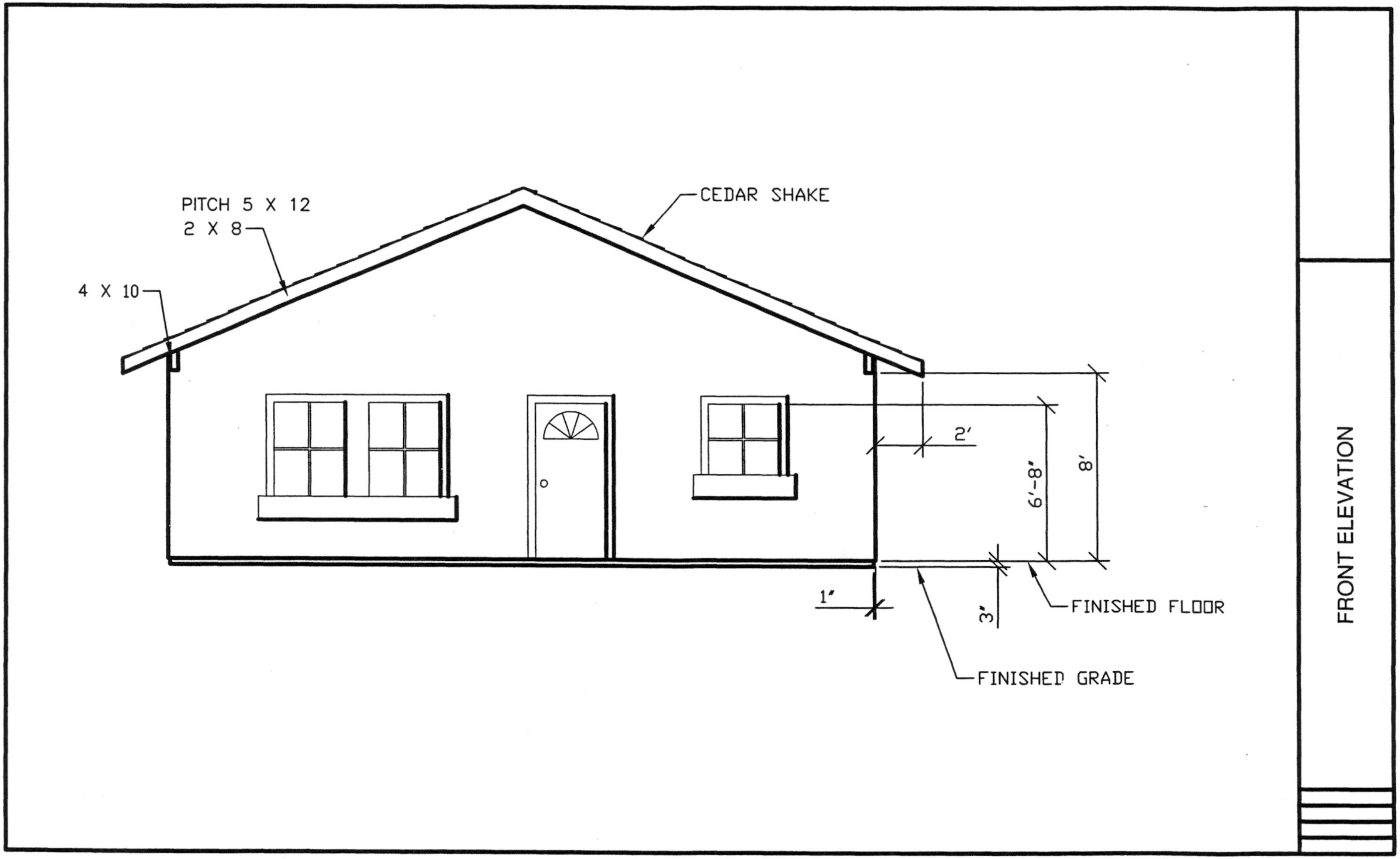

CEDAR SHAKE

PITCH 5 X 12

2 X 8

4 X 10

8'

9'-8'

2'

3'

1'

FINISHED FLOOR

FINISHED GRADE

EXERCISE 158-4

SCALE = 3/8" = 1'

LOT 2 BLOCK A
PINE HILLS SUBDIVISION
SANTA ANA, CALIFORNIA

INSTRUCTIONS:
1. Open **My FEET-INCHES Setup**
2. Select the **QTR EQUALS FOOT** tab.
3. Draw the Front Elevation shown above. Refer to 158-2 for dimensions.
4. Refer to "How to draw shingles" on the next page.
5. Use appropriate layers but colors are your choice.
6. Design is your choice.
7. Fill in the information in the title block area.
8. Adjust the scale of Model Space to **3/8"=1'**
9. Save as: **EX-158-4** and Plot using Page Setup: **24 x18 ALL BLACK**

HOW TO DRAW SHINGLES

The following is an example of how to draw shingles using the ARRAY command.

1. Draw the Roof line **HORIZONTAL.**

_____

2. Draw 2 Shingles as shown below.

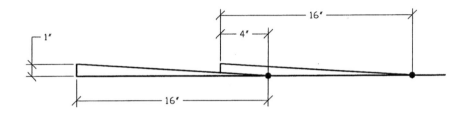

3. ARRAY the 2nd shingle as follows:
    a. Select MODIFY / ARRAY / RECTANGULAR
    b. Select the 2nd Shingle ONLY
    c. Rows = 1   Columns = 17
    d. Distance between columns:
        Snap to 1st point, then snap to 2nd point as shown below.

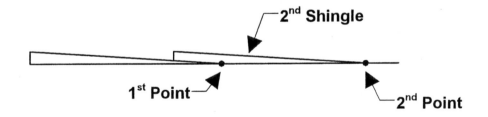

4. ROTATE the Roof Line and the Shingles 23 degrees.

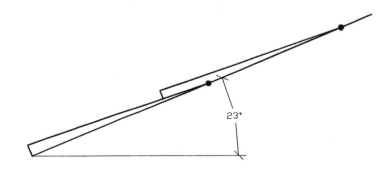

ARCH-12

EXERCISE 158-5
<u>WALL DETAIL</u>

The following exercise requires that you create a new layer with a new linetype.

A. Open **My FEET-INCHES Setup**

B. Select the **24 X 18-FULL** tab.

C. Adjust the scale to : 1" = 1' and **Lock** the Viewport

D. Create a new layer
   Name = Insulation   Color = magenta   Linetype = batting

   Now experiment drawing the insulation as follows:
   1. Select the Layer Insulation
   2. Select Draw / Line
   3. Draw a line

E. Change the **Linetype scale** as follows:
   1. At the command line type: **LTS <enter>**
   2. Type: **.35 <enter>**
   (This will scale the "batting" linetype to match the width of the 2 X 4)

F. Draw the Wall Detail on the next page.
   1. Include the leader call outs.
   2. Use Hatch Pattern "AR-CONC" for footing   Scale = .5
   3. Use User defined for the Finished and Subfloor.
      Angle =45 and 135   Spacing = 3"
   4. Layers and colors, your choice.

G. Save as **158-5**
H. Plot using Page Set up: **24 X 18 ALL BLACK**

EXERCISE 158-5

SCALE = 1" = 1'

2 X 6 RAFTERS AT 16" CENTERS

2 X 6 CEILING JOIST AT 16" CENTERS

(2) 2 X 4 TOP PLATES

3/4" INTERIOR WALL AND CEILING COVER

4" BLANKET INSULATION

1 X 4 BASE BOARD

FINISHED FLOOR

BUILDING PAPER BETWEEN FLOOR LAYERS

1" SUBFLOOR

2 X 8 FLOOR JOIST AT 16" CENTERS

2 X 6 SILL

8" WD X 30" DP FOOTING

12" WD X 6" DP FOOTING

3/4" EXTERIOR COVERING

3/4" SHEATHING

BUILDING PAPER

2 X 4 PLATE

10"

12

7

8'

ELECTRO

MECH

ELECTRO-MECHANICAL SYMBOL LIBRARY

When you are using a CAD system, you should make an effort to ONLY draw an object once. If you have to duplicate the object, use a command such as: Copy, Array, Mirror or Block. Remember, this will make drawing with CAD more efficient.

In the following exercise, you will create a file full of electronic symbols that you consistently use when creating an architectural drawing. You will create them once and then merely drag and drop them, from the DesignCenter, when needed. This will save you many hours in the future.

Save this library file as **Library** so it will be easy to find when using the DesignCenter or create a Library Palette.

1. Open **My Decimal Setup**

2. Add the following layers to your **MY Decimal set up** drawing.

CIRCUIT	GREEN	CONTINUOUS
CIRCUIT2	RED	CONTINUOUS
CORNERMARK	9	CONTINUOUS
DESIGNATOR	BLUE	CONTINUOUS
MISC	CYAN	CONTINUOUS
PADS	GREEN	CONTINUOUS
PCB	WHITE	CONTINUOUS
COMPONENTS	RED	CONTINUOUS

3. **Save the My Decimal Set up.**

4. Select the **11 x 17 (1 to 1)** tab.

5. Draw each of the Symbol objects, shown on the next page, actual size. Do not scale the objects.

6. Create an individual Block for each one using the **BLOCK** command.

7. Save this drawing as: **Library**

8. Plot a drawing of your library symbols for reference. The format is your choice.

9. Plot using Page Set up: **11 x 17 (1 to 1) All Black**

SYMBOLS

INSTRUCTIONS

1

LAYER = COMPONENTS

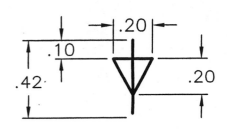

ANTENNA

2

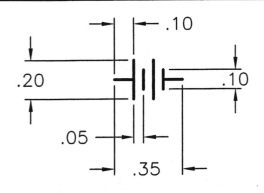

BATTERY

3

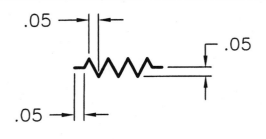

RESISTOR

4

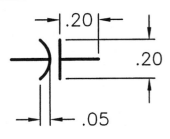

CAPACITOR

5

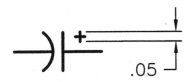

CAPACITOR-POLARIZED

ELECT-3

SYMBOLS

INSTRUCTIONS

6

CAPACITOR-VARIABLE

LAYER = COMPONENTS
SAME AS SYMBOL 4
USE POLYLINE FOR ARROW

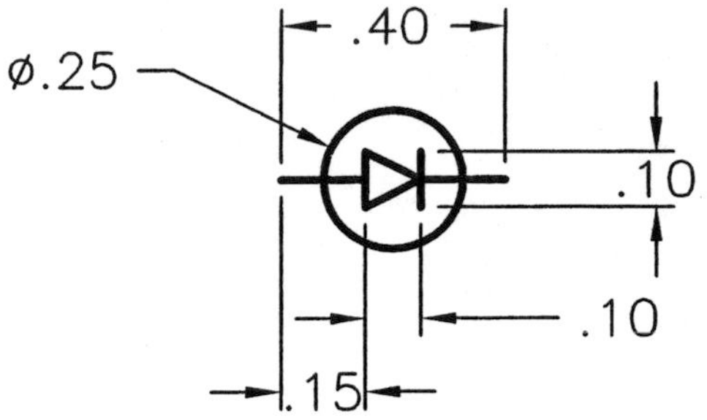

7

Ø.25

.40

.10

.10

.15

DIODE

LAYER = COMPONENTS

8

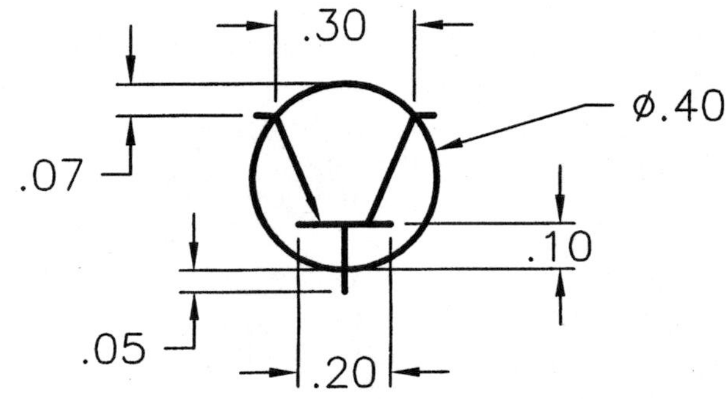

.30

Ø.40

.07

.10

.05

.20

TRANSISTOR-PNP

9

TRANSISTOR-NPN

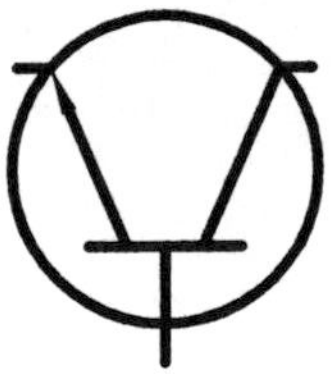

10

Ø.05

.20

SWITCH

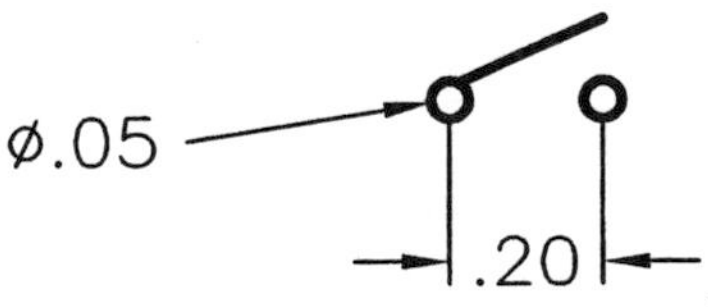

ELECT—4

SYMBOLS

INSTRUCTIONS

11

LAYER = COMPONENTS

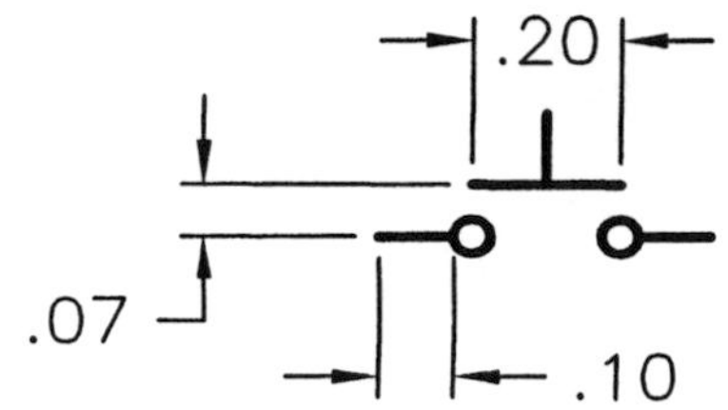

→|.20|←

.07

.10

SWITCH-PUSH BUTTON

12

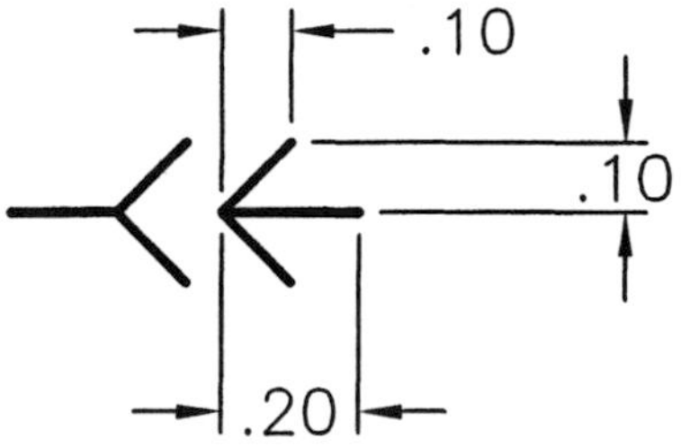

.10

.10

.20

PLUG

13

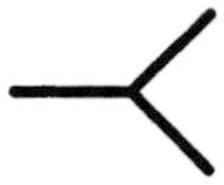

JACK - FEMALE

14

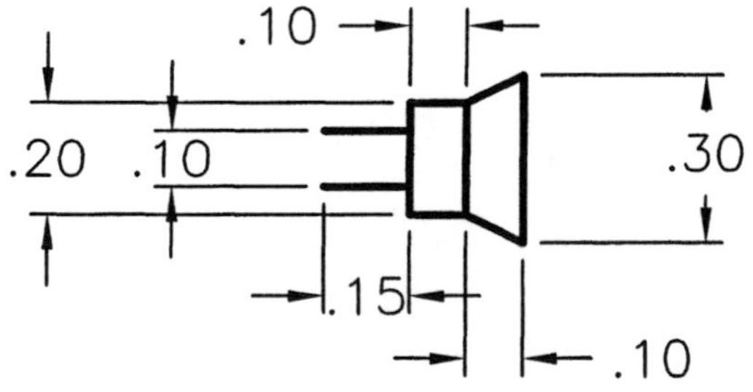

.10

.20 .10

.30

.15

.10

SPEAKER

15

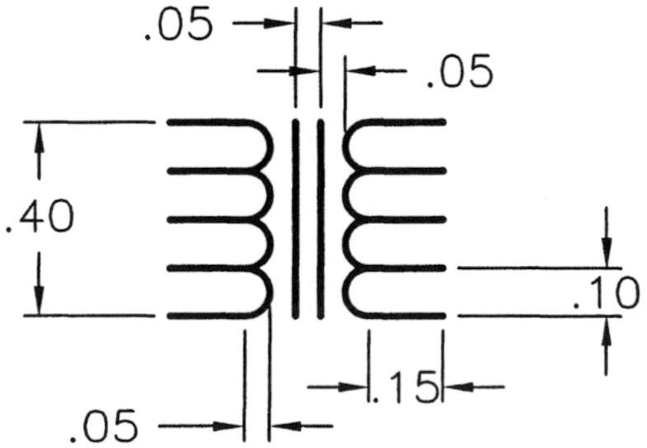

.05

.05

.40

.10

.15

.05

CAPACITOR-POLARIZED

SYMBOLS

INSTRUCTIONS

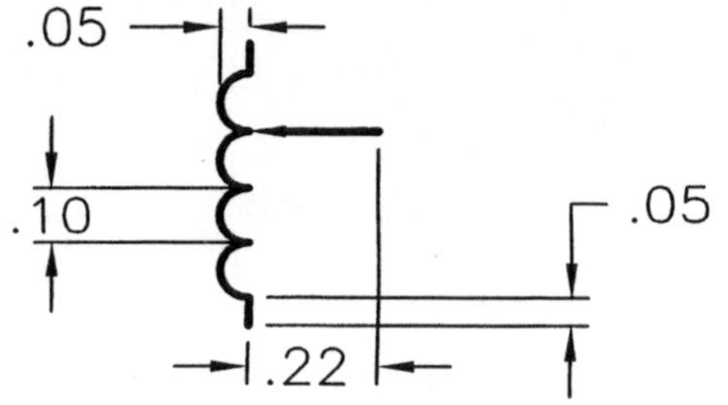

16

.05

.10

.05

.22

INDUCTOR - VARIABLE

LAYER = COMPONENTS

17

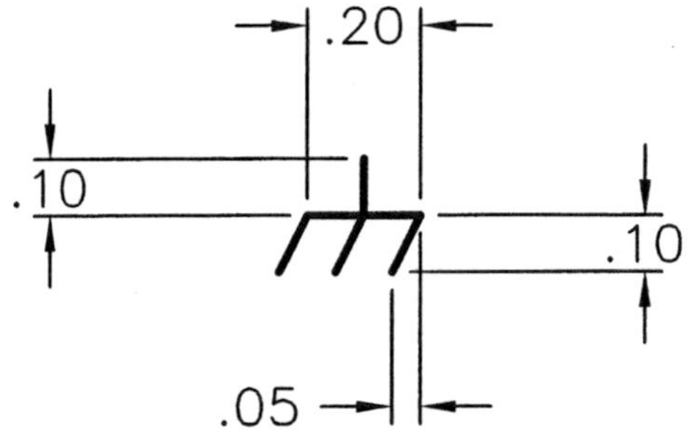

.20

.10

.10

.05

CHASSIS GROUND

18

.40

.10

.10

RESISTOR (1/4W)

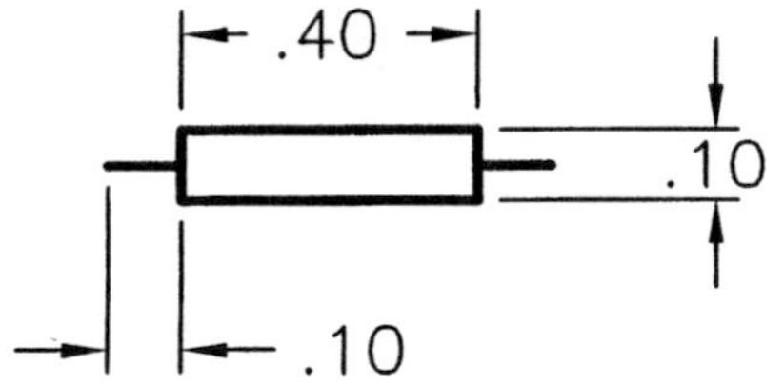

19

.40

.15

.10

RESISTOR (1/2W)

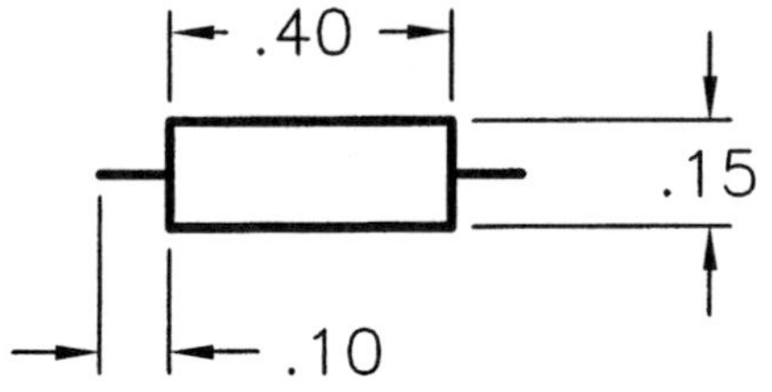

20

.70

.20

.15

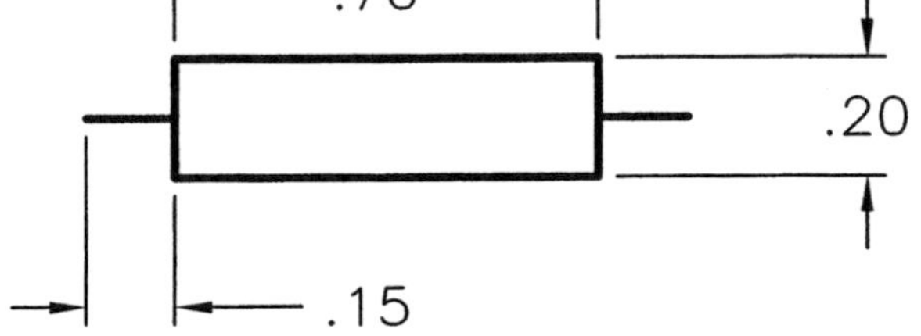

RESISTOR (2W)

ELECT-6

SYMBOLS

INSTRUCTIONS

21

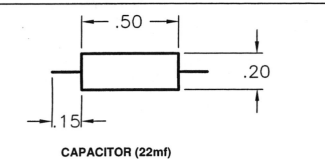

CAPACITOR (22mf)

22

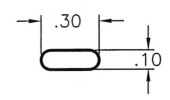

CAPACITOR (.01mf)

23

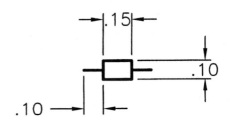

DIODE (IN914)

24

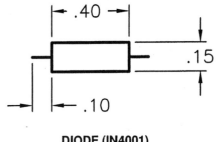

DIODE (IN4001)

25

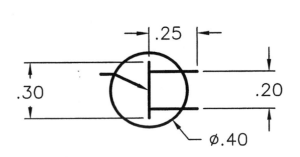

2N2646 UJT (USE FOR SCHEMATIC ONLY)

LAYER = COMPONENTS

ELECT-7

SYMBOLS

INSTRUCTIONS

26

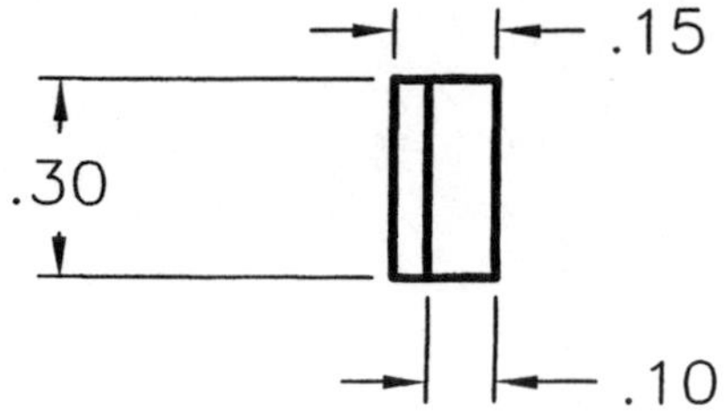

TIP29A

27

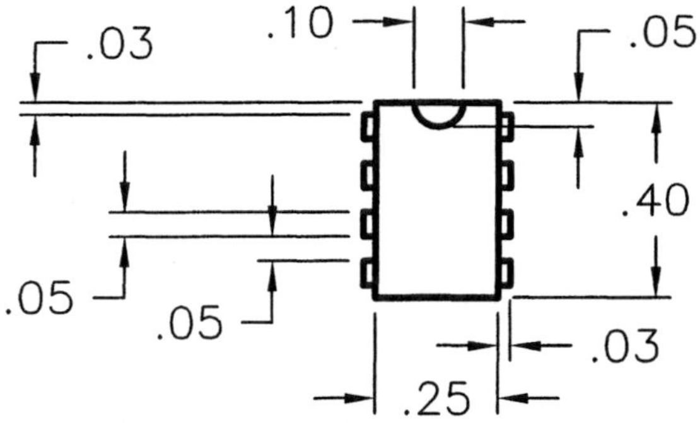

TIMER-NE555

28

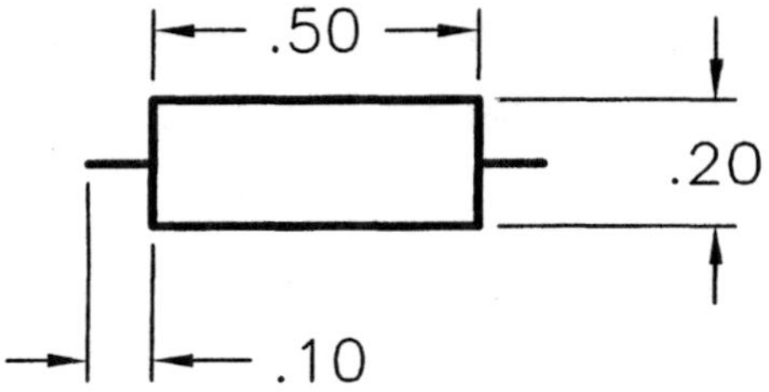

CAPACITOR

29

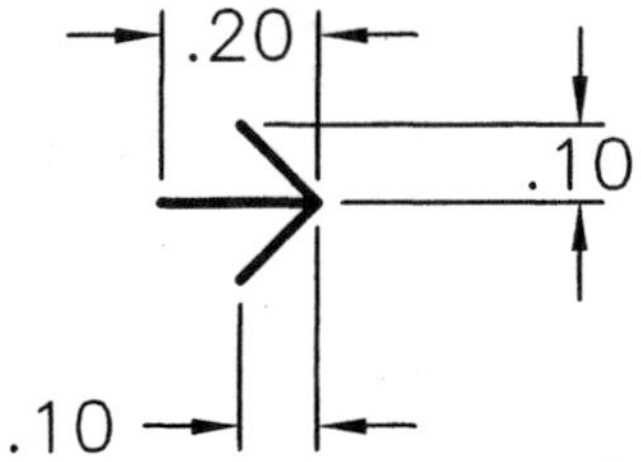

PLUG-MALE

30

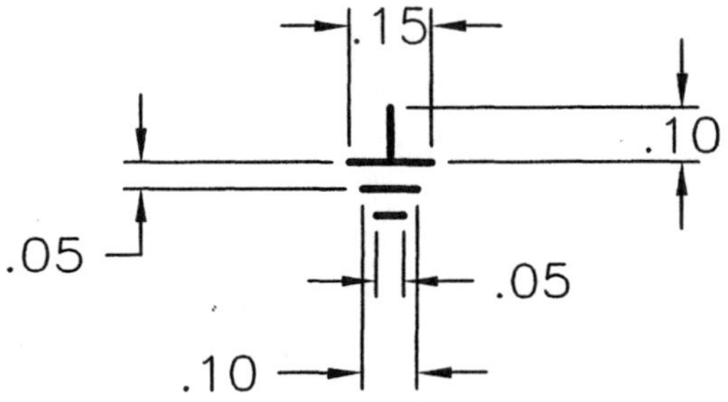

EARTH GROUND

LAYER = COMPONENTS

ELECT-8

SYMBOLS		INSTRUCTIONS

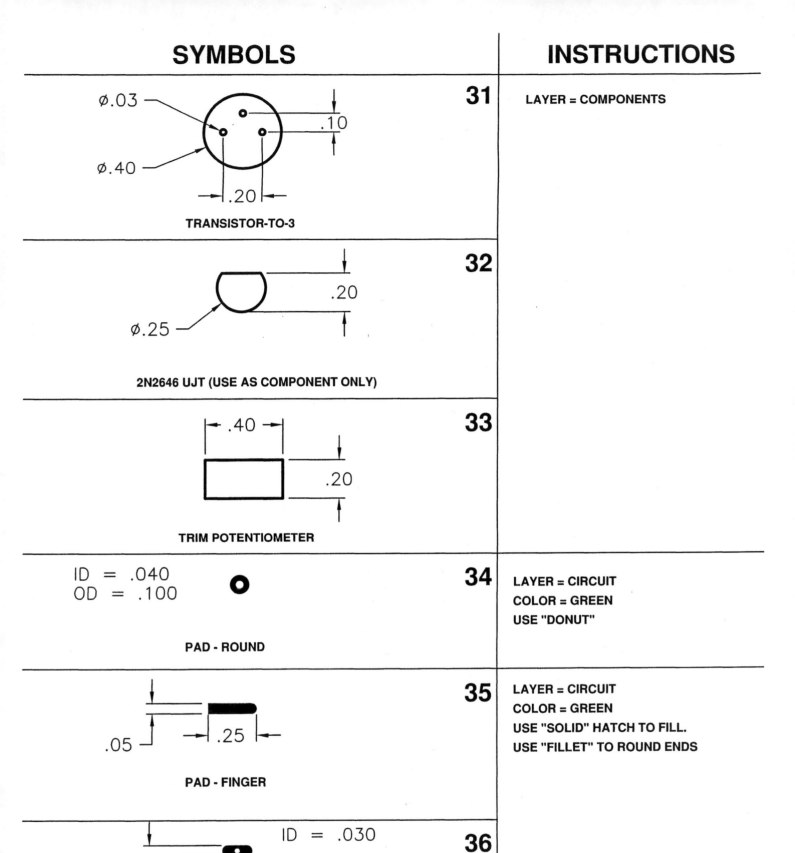

31

LAYER = COMPONENTS

Ø.03
Ø.40
.10
.20

TRANSISTOR-TO-3

32

Ø.25
.20

2N2646 UJT (USE AS COMPONENT ONLY)

33

.40
.20

TRIM POTENTIOMETER

34

ID = .040
OD = .100

PAD - ROUND

LAYER = CIRCUIT
COLOR = GREEN
USE "DONUT"

35

.05
.25

PAD - FINGER

LAYER = CIRCUIT
COLOR = GREEN
USE "SOLID" HATCH TO FILL.
USE "FILLET" TO ROUND ENDS

36

ID = .030
.06
.150

PAD - RECTANGULAR

ELECT-9

EXERCISE 156-1

BLOCK DIAGRAM

RF
AMPLIFIER

OSCILLATOR
MIXER

IF
AMPLIFIER

DETECTOR

AUDIO
AMPLIFIER

INSTRUCTIONS:

1. Open **MY DECIMAL SET UP**
2. Use Layer: **TEXT LIGHT** for text. Use Layer **OBJECT** for objects and symbols.
3. Size and proportions are your choice.
4. Save as: **EX-156-1**
5. Plot using Page Set up: **11 X 17 (1 to 1) ALL BLACK**

ELECT-10

EXERCISE 156-2

The following is an example of a how you might construct a schematic.

A. Open **MY DECIMAL SETUP**

B. Select the **11 X 17 (1 to 1)** tab.

C. Set   GRID = .100   SNAP = .100

D.   Draw the schematic in Model Space.
    The size is not critical but maintain good proportions and drawing balance.

SUGGESTION:

1. Use layer **CONSTRUCTION** to roughly layout the circuit lines
2. Change to Layer **Circuit** before inserting symbols.
3. **Insert** symbols and locate as shown. (Drag and drop from the DesignCenter, Library drawing
4. Draw the **actual** schematic lines on top of the Construction Lines skipping over the symbols. Use layer **circuit2**
5. Now turn **off** layer **Construction**

(Now you don't have to trim the lines under the symbols.)

E.   Draw the solder points using **Donuts**   .00 I.D. and .100 O.D.
     Use layer **Circuit**

F.   Add the designators, use layer **Designator.**
    Text Ht. = .125

G.   Add the Parts List (In Paperspace) Rows = .25   Text Ht. = .125

H.   Edit the title block

I.   Save as **EX-156-2**

J.   Plot using Page Set up: **11 X 17 (1 to 1) All Black**

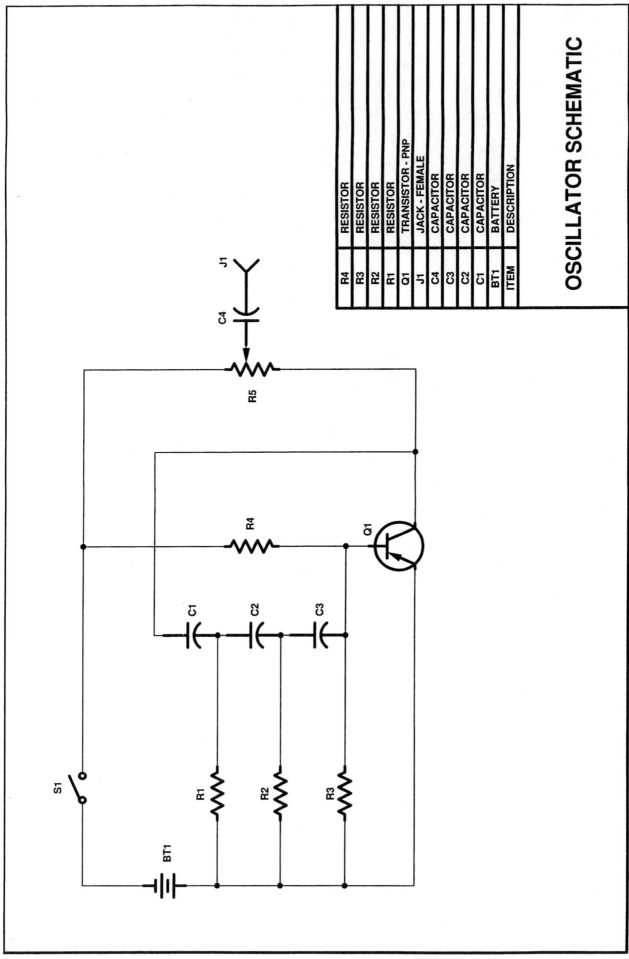

ITEM	DESCRIPTION
R4	RESISTOR
R3	RESISTOR
R2	RESISTOR
R1	RESISTOR
Q1	TRANSISTOR - PNP
J1	JACK - FEMALE
C4	CAPACITOR
C3	CAPACITOR
C2	CAPACITOR
C1	CAPACITOR
BT1	BATTERY

OSCILLATOR SCHEMATIC

EXERCISE 156-2

EXERCISE 156-3

This drawing is only a template. It will be used as a template for the following 4 drawings. If you follow the instructions and draw this template correctly, the next 4 drawings will be very easy.

A. Open **MY DECIMAL SETUP**

B. Select the **11 X 17 (1 to 1)** tab and unlock the viewport.

C. Adjust the scale of model space to 2 : 1

D. Set   GRID = .100   SNAP = .100

E. Draw the board outline (on layer PCB) Full Scale (2.50 x 2.00).
(Note: it will appear larger because you adjusted the scale of model space to 2 :1)
Printed Circuit Boards are generally drawn 2, 10 or even 100 times larger than their actual size.

F. Draw the **Circuit** on layer **CONSTRUCTION.**

G. **INSERT** symbols 19, 28, and   31 on Layer **Components** (Refer to Library)

H. Draw the designators on Layer **DESIGNATOR**. Text Ht = .060

I. Draw dimensions, in paper space, on Layer **DIMENSION**.

IMPORTANT: Make sure that True Associative Dimensioning is On:

Follow the instructions below:
1. Type **Dimassoc <enter>**
2. Type **2 <enter>**

J. **Edit** the title block

K.   Save as **EX-156-3 (Do not plot)**

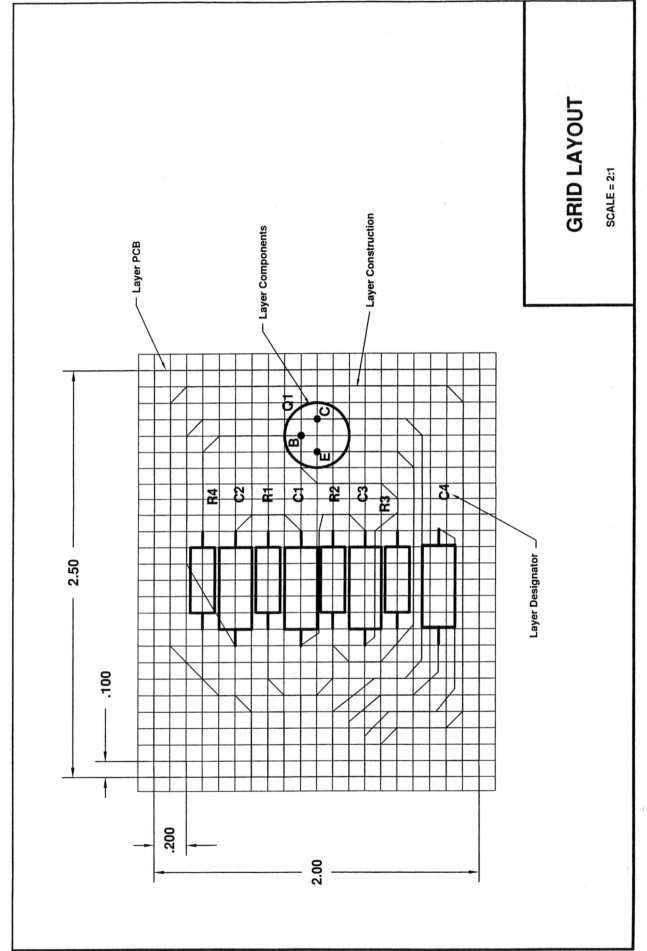

GRID LAYOUT

SCALE = 2:1

EXERCISE 156-3

Layer PCB

Layer Components

Layer Construction

Layer Designator

2.50

.100

.200

2.00

Note: Do not draw the grid. It is there only to help you estimate positions.

ELECT-14

EXERCISE 156-4

The following is an example of a how you might illustrate the ARTWORK for the circuit on the PCB.

A. Open **EX-156-3**

B. Turn **OFF** Layers **DESIGNATOR, DIMENSION** and **COMPONENTS**.
    (Remember, if you were not careful when you created EX-156-3 the wrong objects may disappear. You may have to move objects to the correct layer)

C. Draw the heavy lines outside the corners of the PCB. (This defines the edges of the board)
    Use layer Cornermark.
    Use Polyline, width .100

Note: Consider using Offset to create a guideline, .05 from the edge of the board, for the polyline. Then just snap to the intersections.

D. **Insert** symbols **34** and **35** on layer **CIRCUIT.**

E. Draw the circuit lines **on top** of the construction lines.
    Use Layer Circuit
    Use Polyline, width .025

F. Turn **OFF** Layers, **CONSTRUCTION** and **PCB.**

G. Save as **EX-156-4**

H. **Plot** using Page Setup: **11 X 17 (1 to 1) All Black**

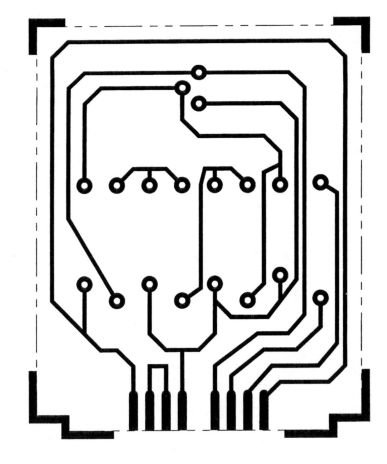

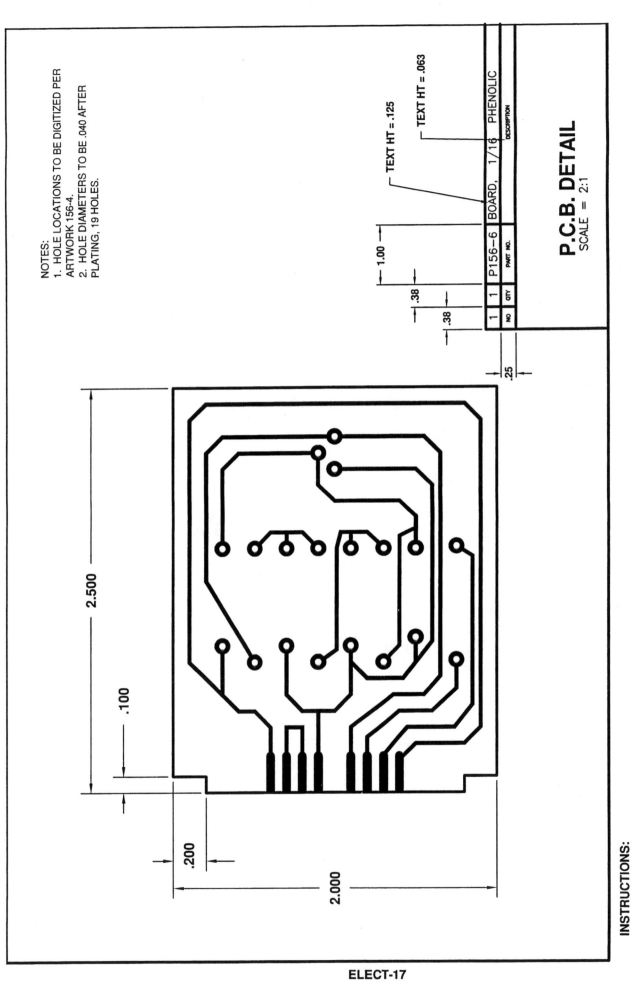

NOTES:
1. HOLE LOCATIONS TO BE DIGITIZED PER ARTWORK 156-4.
2. HOLE DIAMETERS TO BE .040 AFTER PLATING, 19 HOLES.

TEXT HT = .125

TEXT HT = .063

NO	QTY	PART NO.	DESCRIPTION
1	1	P156-6	BOARD, 1/16 PHENOLIC

1.00

.38

.38

.25

P.C.B. DETAIL
SCALE = 2:1

2.500

.100

.200

2.000

EXERCISE 156-5

INSTRUCTIONS:
1. Open EX-156-4
2. Freeze Layer Cornermark and Thaw Layers PCB and Dimension
3. Draw the Parts List in Paperspace. Use Layer: Border and Text Light.
4. Add the notes on Layer Text Light. Ht = .125
5. Save as EX-156-5 and Plot using Page Set up: 11 X 17 (1 to 1) All Black.

ELECT-17

EXERCISE 156-6

The following is an example of a how you might illustrate the **ARTWORK** for the **COMPONENTS**.

A. Open **EX-156-3**

B. Turn **OFF** Layers **CONSTRUCTION and DIMENSION..**
    (Remember, if you were not careful when you created EX-156-3 the wrong objects may disappear. You may have to move objects to the correct layer)

C. **MIRROR** the entire board and its components.

    WHY? A printed circuit board has the circuit on one side and the components on the other. We need to show the opposite side of the board.

    First, lets learn a new command.
    1. At the command line type: **MIRRTEXT <enter>**
    2. Type: **0 <enter>**   (0 means OFF,   1 means ON)

    The **MIRRTEXT** command controls whether the TEXT will mirror or not. It will change positions with the object but you can control the "Right Reading". In this case, we do not want the text to be shown reversed, so we set the MIRRTEXT command to "0" OFF.

D. ALIGN the designators.

E. Add the Note, in paper space, on layer TEXT LIGHT.   Text ht = .125

F. Save as: **EX-156-6**

G. **Plot** using Page Setup: **11 X 17 (1 to 1) All Black**

EXERCISE 156-6

COMPONENT ARTWORK

NOTES:
1. SILKSCREEN TO BE PAINT, WHITE.

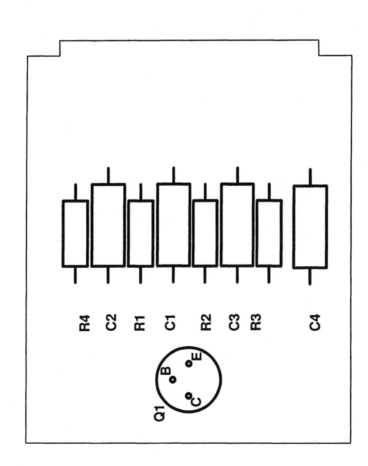

R4 C2 R1 C1 R2 C3 R3 C4

Q1

EXERCISE 156-7

ITEM	QTY	PART NO.	DESCRIPTION	REF. DESIGNATOR
3	1	Q45666	TRANSISTOR TO-3	Q1
2	4	C11227	CAPACITOR 0.01mf	C1-C4
1	4	R22477	RESISTOR 22K 1/2WATT	R1-R4

PCB ASSEMBLY

SCALE = 2:1

1.00

1.00

.38

.38

.25

TEXT HT. = .125

TEXT HT = .063

R4 C2 R1 C1 R2 C3 R3 C4

Q1

INSTRUCTIONS:
1. Open EX-156-6
2. Make the modification to the drawing. (These are the actual components shown in place)
3. Delete the Note.
4. Draw the Parts list, in paperspace.
5. Save as EX-156-7 and Plot using Page Setup: 11 X 17 (1 to 1) All Black.

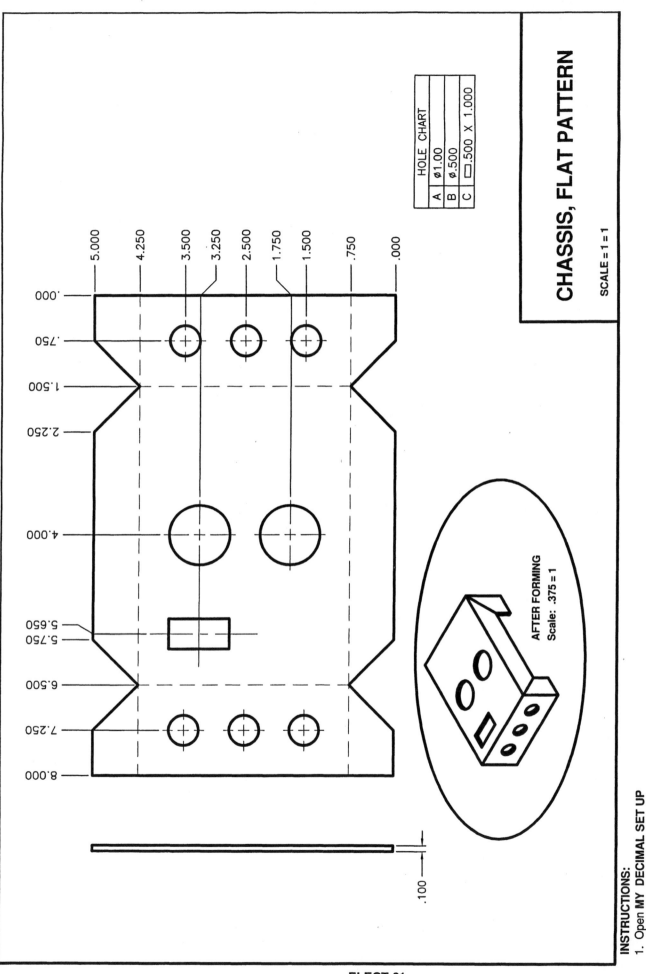

HOLE CHART

A	⌀1.00	
B	⌀.500	
C	⬜.500 X 1.000	

CHASSIS, FLAT PATTERN

SCALE = 1 = 1

EXERCISE 156-8

AFTER FORMING
Scale: .375 = 1

INSTRUCTIONS:

1. Open **MY DECIMAL SET UP**
2. Adjust model space scale to 1 : 1
3. Draw the "Flat Pattern" as shown, FULL SCALE.
4. Dimension, in paper space, using ORDINATE dimensioning.
5. Draw the "Isometric" view to illustrate the "After forming" appearance. (Cut another Viewport)
6. Draw the "Hole chart" in paper space. Size is your choice.
7. Save as: **EX-156-8** and Plot using Page Set up: **11 X 17 (1 to 1) All Black**

ELECT-21

NOTES:

MECHANICAL

MECHANICAL SYMBOL LIBRARY

When you are using a CAD system, you should make an effort to ONLY draw an object once. If you have to duplicate the object, use a command such as: Copy, Array, Mirror or Block. Remember, this will make drawing with CAD more efficient.

In the following exercise, you will create two mechanical symbols that you use frequently. You will create them once and then merely drag and drop them, from the DesignCenter, when needed. This will save you many hours in the future.

Save this library file as **Library** so it will be easy to find when using the DesignCenter or create a Library Palette.

1. Open **My Decimal Setup**

2. Select the **11 X 17 (1 to 1)** tab.

3. Draw each of the Symbol objects, shown on the next page, actual size. Do not scale.

4. Create an individual Block for each one using the **BLOCK** command.

5. Save this drawing as: **Library**

6. Plot a drawing of your library symbols for reference. The format is your choice.

7. Plot using Page Set up: **11 X 17 (1 to 1) All Black**

SYMBOL

INSTRUCTIONS

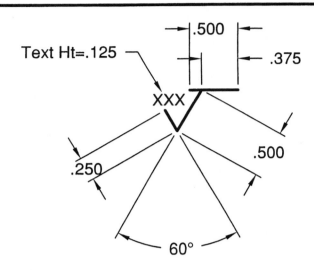

Text Ht=.125

.500

.375

XXX

.250

.500

60°

FINISH MARK

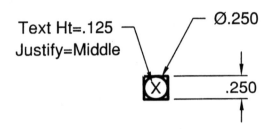

Text Ht=.125
Justify=Middle

Ø.250

.250

X

NOTE IDENTIFIER

EXERCISE 157-1

ECCENTRIC HUB

SCALE: 1 = 1

Ø5.00

Ø2.375

125

⌖ Ø.028 Ⓜ A B

.50 ECC

.50

1.500

1.500

8X Ø.570 / .563

⌖ Ø.028 Ⓜ A B

2X Ø.50-13UNC

Ø 6.00

Ø 4.000 / 3.995

125

∥ .002 A B

250

.002 A

125

.50

1.00

▱ .002 A

NOTE: UNLESS OTHERWISE SPECIFIED:

1. DIMENSIONING AND TOLERANCING IAW ANSI Y14.5.
2. BREAK ALL SHARP CORNERS.
3. MATERIAL: 1045 CRS.

INSTRUCTIONS:

1. Open **MY DECIMAL SETUP**
2. Select the **24 X 18 (1 to 1)** tab.
3. Draw ECCENTRIC HUB above, in model space, full scale.
4. Dimension as shown. Set the Tolerance Text Ht to .125.
5. Insert the symbols, on the Symbols Layer, using DesignCenter.
6. Fill in the information in the title block area
7. Save as: **EX-157-1** and Plot using Page Setup: **24 X 18 (1 to 1) All Black**

MECH-4

SPLINE HUB

SCALE: 1 = 1

EXERCISE 157-2

6X ⌀ .312/.310

⌀ .756/.750   ⟂ .0015 A B

④

⌀ 3.375

1.56-16 UN-2A   ↗ .005 B

⌀ 1.250/1.248   ↗ .005 B

⌀ 1.756/1.750   ↗ .003 B

⌀ 1.44/1.42

⌀ 4.00/3.98

.26/.24

// .0025 A

.380/.370

.380/.370

.380/.370

1.005/1.000

1.40/1.35

1.626/1.620

⌺ .001 A

250

250

250

250

250

C

2.26/2.23

NOTE: UNLESS OTHERWISE SPECIFIED:

1. DIMENSIONING AND TOLERANCING IAW
   ANSI Y14.5.
2. BREAK ALL SHARP CORNERS.
3. MATERIAL: 1045 CRS.
④ RUBBER STAMP WITH .12 BLACK CHARACTERS
   PER MIL-STD-130.
5. ALL RADII AND FILLETS ARE R.09

INSTRUCTIONS:

1. Open **MY DECIMAL SETUP**
2. Select the **24 X 18 (1 to 1)** tab.
3. Draw **SPLINE HUB** above, in model space, full scale.   Use Hatch Pattern = ANSI 31
4. Dimension as shown. Set the Tolerance Text Ht to .125.
5. Insert the symbols, on the Symbols Layer, using DesignCenter.
6. Fill in the information in the title block area
7. Save as: **EX-157-2** and Plot using Page Setup: **24 X 18 (1 to 1) All Black**

MECH-5

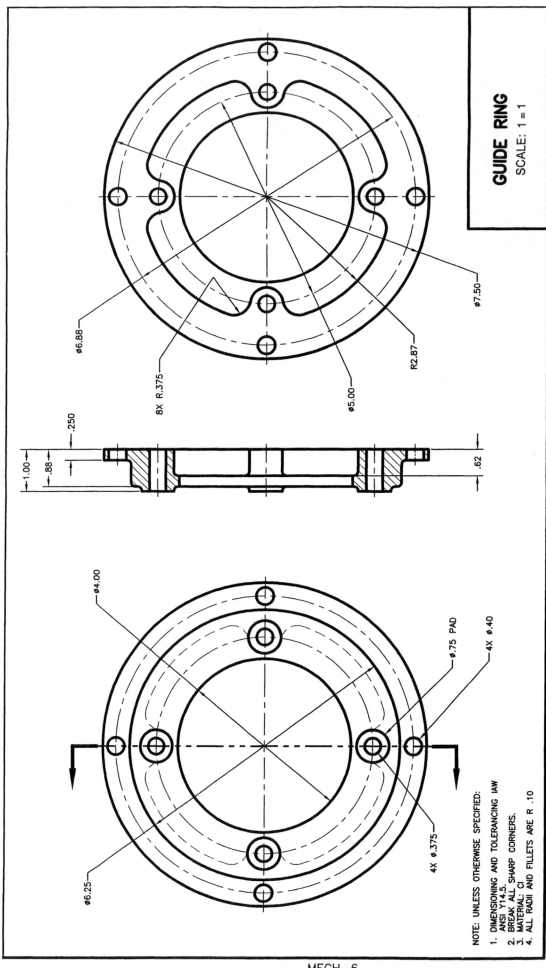

GUIDE RING

SCALE: 1 = 1

EXERCISE 157-3

Ø6.88

.250

1.00

.88

8X R.375

Ø4.00

Ø.75 PAD

4X Ø.40

Ø5.00

R2.87

Ø7.50

.62

4X .375

Ø6.25

NOTE: UNLESS OTHERWISE SPECIFIED:

1. DIMENSIONING AND TOLERANCING IAW
   ANSI Y14.5.
2. BREAK ALL SHARP CORNERS.
3. MATERIAL CI
4. ALL RADII AND FILLETS ARE R .10

INSTRUCTIONS:

Open **MY DECIMAL SETUP**

1. Open **MY DECIMAL SETUP**
2. Select the **24 X 18 (1 to 1)** tab.
3. Draw GUIDE RING above, in model space, full scale. Use POLYLINE for the Section Line. (width = .04)
4. Dimension as shown.
5. Hatch Pattern = ANSI 31.
6. Fill in the information in the title block area
7. Save as: **EX-157-3** and Plot using Page Setup: **24 X 18 (1 to 1) All Black**

MECH−6

APPENDIX A
Add a Printer / Plotter

The following are step by step instructions on how to configure AutoCAD for your printer or plotter. These instructions assume you are a single system user. If you are networked or need more detailed information, please refer to your AutoCAD users guide.

A. Select **File / Plotter Manager**
B. Select "Add-a-Plotter" Wizard

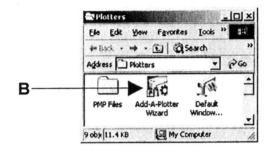

The following dialog boxes will appears.

C. Select the **"Next"** button.

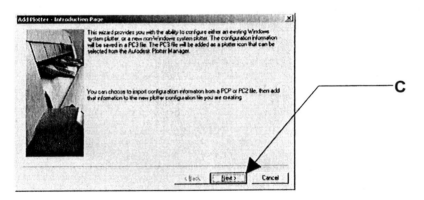

D. Select **"My Computer"** then **Next**.

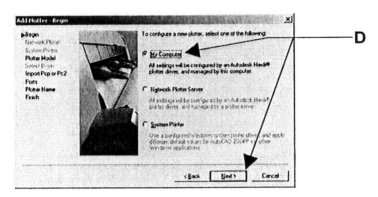

E. Select the **Manufacturer** and the specific **Model** desired then **Next**.

(If you have a disk with the specific driver information, put the disk in the disk drive and select "Have disk" button then follow instructions.)

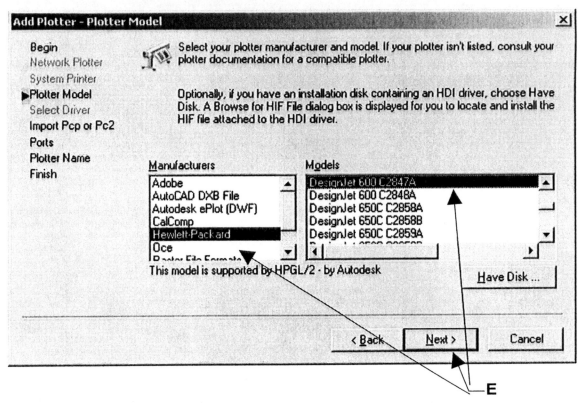

F. Read the next screen and select **"Continue or Exit"**

G. <u>Do not</u> select the Import File button, just select **Next**

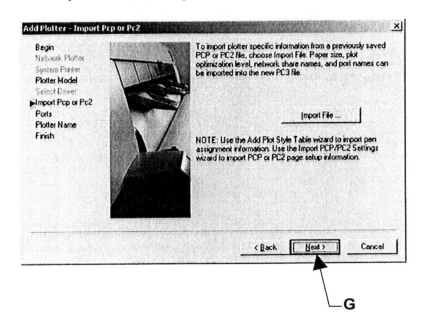

H. Select <u>LPT1</u> or <u>COM1</u> or <u>COM2</u>, then **NEXT.**

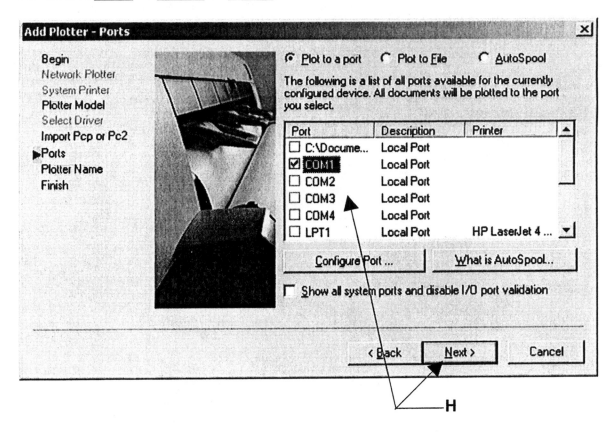

I. Accept the Plotter Name or type a different name, then select **NEXT.**

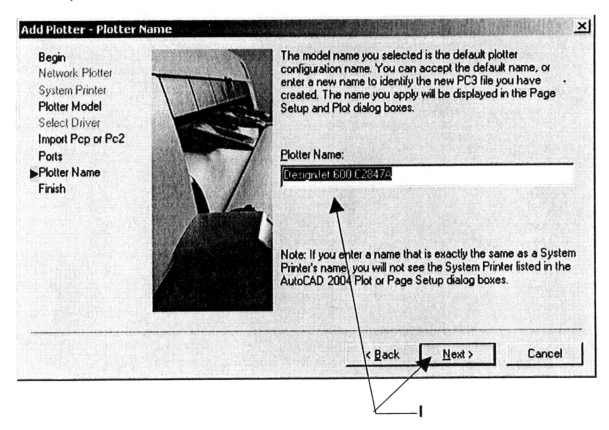

J. Select **FINISH**

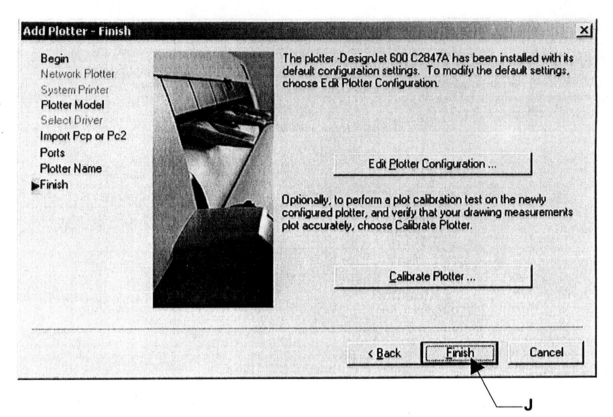

K. Now check the **File / Plotter Manager.** Is the Printer / Plotter there?

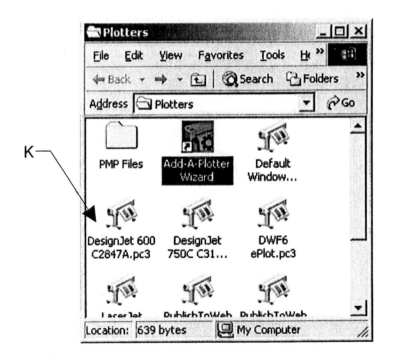

APPENDIX B
Printing In Color Or Black <u>with Lineweights</u>

Even though you draw with colors, you may print the drawing in <u>color</u> or <u>black</u>.
To print with multiple colors select the Plot Style table "None".
To print with black only, you must create a "<u>color-dependent plot style table</u>".
(**.ctb** file.)

HOW TO CREATE A "COLOR-DEPENDENT PLOT STYLE TABLE"

1. Select **File / Plot**

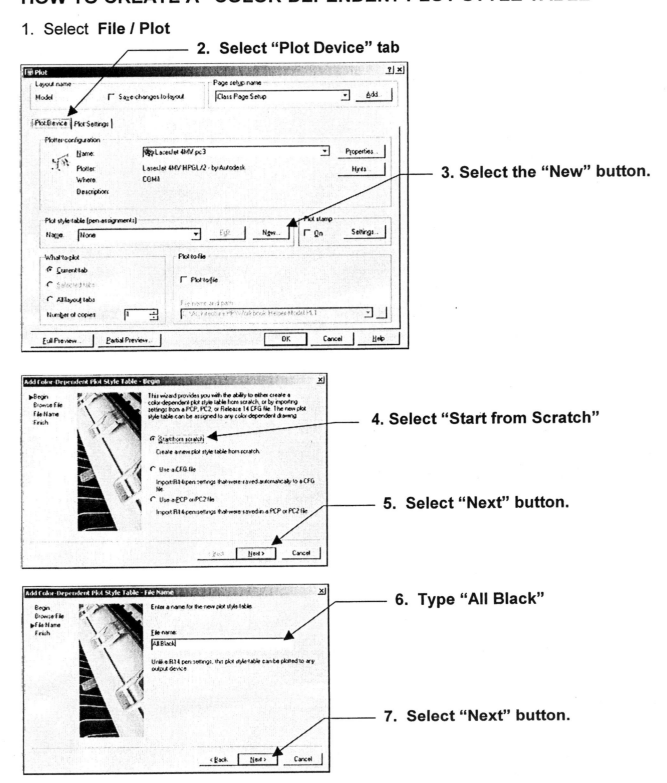

2. Select "Plot Device" tab

3. Select the "New" button.

4. Select "Start from Scratch"

5. Select "Next" button.

6. Type "All Black"

7. Select "Next" button.

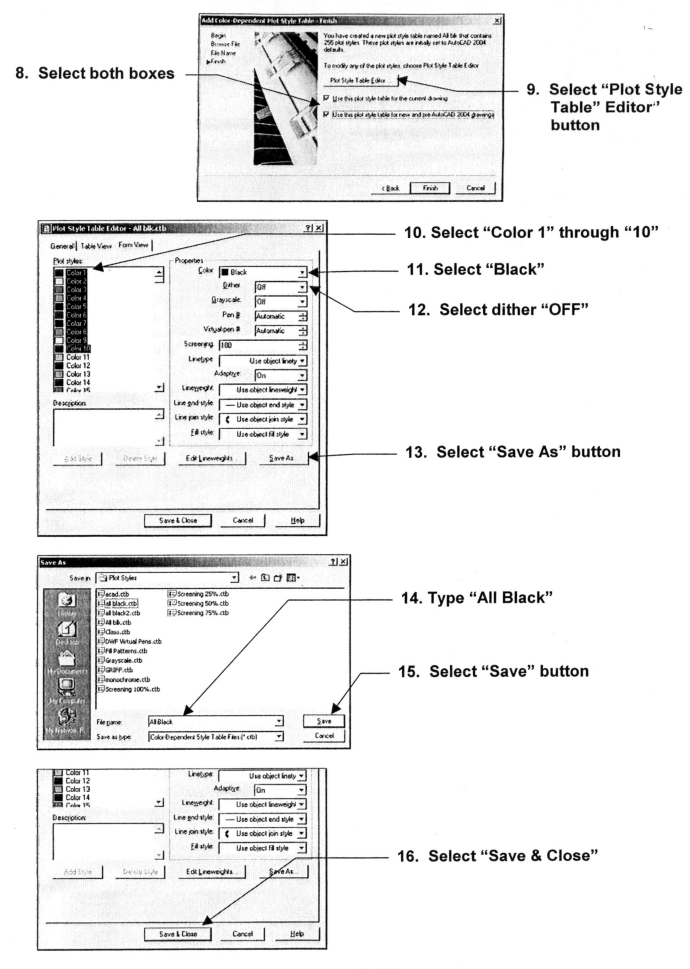

8. Select both boxes

9. Select "Plot Style Table" Editor" button

10. Select "Color 1" through "10"

11. Select "Black"

12. Select dither "OFF"

13. Select "Save As" button

14. Type "All Black"

15. Select "Save" button

16. Select "Save & Close"

Appendix-B2

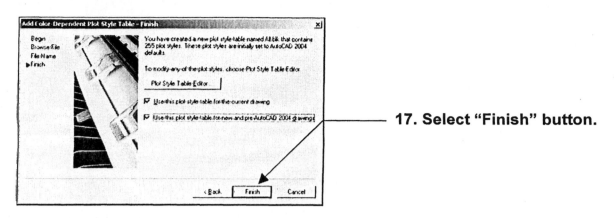

17. Select "Finish" button.

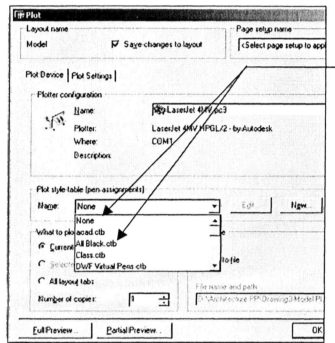

Now you have (2) .ctb files:

"None" **to be used when you want to print with color.**

"All Black" **to be used when you want to print Black only.**

NOTE:

If you have Lineweights assigned within the drawing:
The steps shown above assists you in creating a "Color-Dependent Plot Style table" to be used if you have lineweights assigned within the drawing. The color will be controlled by the Plot Style selected and the Lineweights plotted will be controlled by the lineweights assigned within the drawing.

If you do not have Lineweights assigned within the drawing:
You do not have to assign Lineweights within the drawing to plot with Lineweights. You may assign lineweights within a Plot Style. The steps on the next page will assist you in creating a "Color-Dependent Plot Style table" to assign Lineweights to colors.

NOTES:

APPENDIX C
How To Create A New "<u>Color-Dependent</u> <u>Plot Style</u> <u>Table</u>" To Be Used <u>Without Lineweights</u>

Each color, used within a drawing, can be assigned different plotting properties. Plotting properties are line weight, line type, line end style, joint style, fill pattern, gray scale, screen percentage etc. When you plot your drawing using a specific Plot Style Table, the plotting properties you have assigned will apply to the color of the objects in your drawing.

For example: You could create a Plot Style Table called "XYZ" with the line weight of color red set to .0831 inches. When you plot the drawing you will select the "XYZ" Plot Style Table. All the objects that are red, in the drawing, will have a line weight of .0831 inches on the paper when plotted.

A. Select **FILE / PLOT STYLE MANAGER**

B. Select **"Add-A-Plot Style Table" Wizard**

Add-A-Plot
Style Tab...

The following dialog boxes will appear.

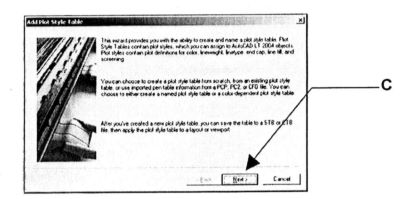

C. Select the **Next** button.

D. Select **"Start from Scratch"** then the **Next** button.

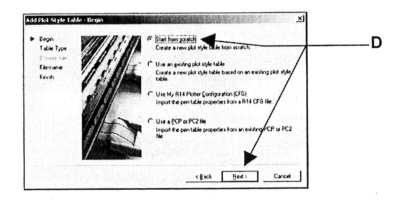

E. Select **"Color-Dependent Plot Style Table"** then the **Next** button.

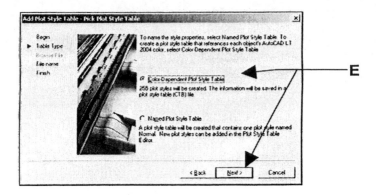

F. Type the new Plot Style Table **name** then select the **Next** button.

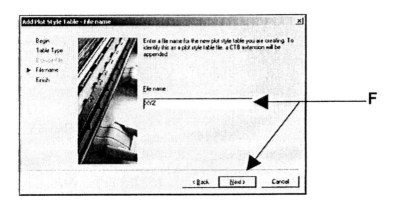

G. Select the **"Plot Style Table Editor"** button.

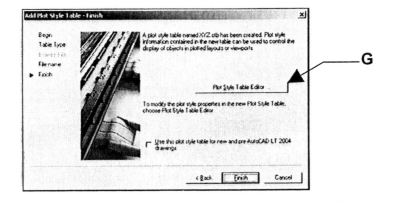

H. Make changes to the **"PROPERTIES"** then select the **Save & Close** button.

1. Select color 1

2. Change color property to Black

3. Change Lineweight to: 2.1100mm or 0.0831 inches

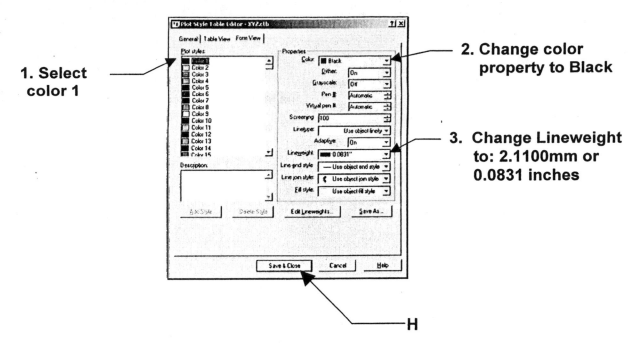

I. Select the **FINISH** button.

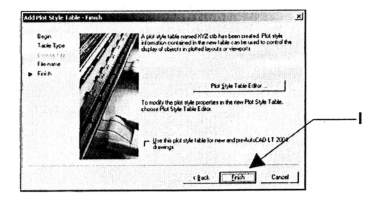

Now select File / Plot.
Check the list of Plot Style tables to see if your new style is listed.

NOTES:

INDEX

Y

Z
Zaxis rotation, 15-5